OT 58
Operator Theory: Advances and Applications
Vol. 58

Birkhäuser Verlag
Basel · Boston · Berlin

Continuous and Discrete Fourier Transforms, Extension Problems and Wiener-Hopf Equations

Edited by

I. Gohberg

1992

Birkhäuser Verlag
Basel · Boston · Berlin

Editor's address:

Prof. I. Gohberg
Raymond and Beverly Sackler Faculty of Exact Sciences
School of Mathematical Sciences
Tel Aviv University
69978 Tel Aviv, Israel

A CIP catalogue record for this book is available from the Library of Congress, Washington D.C., USA

Deutsche Bibliothek Cataloging-in-Publication Data

Continuous and discrete Fourier transforms, extension problems, and Wiener Hopf equations / ed. by I. Gohberg. –
Basel ; Boston ; Berlin : Birkhäuser, 1992
(Operator theory ; Vol.58)
ISBN 3-7643-2809-6 (Basel ...)
ISBN 0-8176-2809-6 (Boston)
NE: Gochberg, Izrail' C. [Hrsg.]; GT

Printed in Germany on acid-free paper, directly from the authors' camera-ready manuscripts
ISBN 3-7643-2809-6
ISBN 0-8176-2809-6

Table of Contents

EDITORIAL INTRODUCTION

The present book is a selection of papers in modern analysis of Fourier transforms and their applications. It consists of seven papers in which the continuous and discrete Fourier transforms appear as subjects of research, or are used as tools in solving other problems. In some papers the interplay between the continuous and discrete Fourier transforms is very important.

The first four papers are based on talks presented at a workshop held in the Department of Mathematics, Maryland University, College Park, September 25 and 26, 1991. (The program of the workshop is included at the end of this volume.) This conference was dedicated to the continuous and discrete Fourier transforms and extension problems.

The last three papers were added later. They fit the material described above and are concerned with Wiener-Hopf equations and extension problems.

The first paper, J.Benedetto, C.Heil, D.Walnut, "Uncertainty principles for time-frequency operators" is dedicated to sampling results for frequency transformations and uncertainty principles for exact Gabor and wavelets frames.

In the second paper, R.E.Ellis, I.Gohberg, D.C.Lay, "Distribution of zeros of matrix-valued continuous analogues of orthogonal polynomials" continuous analogues of matrix valued orthogonal polynomials are analyzed. The number of zeros in the upper half plane of such polynomials is computed in terms of the number of negative eigenvalues of an associated integral operator. Both continuous and discrete Fourier transforms are used here.

In the third paper I.Gohberg, M.A.Kaashoek, "The band extension on the real line as a limit of discrete band extensions, II. The entropy principle" the continuous analogue of the entropy principles is deduced from the maximum entropy principle in the discrete case. The interplay between continuous and discrete Fourier transforms is essential.

The fourth paper, M.Cotlar, C.Sadosky, "Weakly positive matrix measures, generalized Toeplitz forms, and their applications to Hankel and Hilbert transform operators" is a review paper in which a number of important recent results of the same authors, are reviewed and commented upon.

The fifth paper, J.A.Ball, I.Gohberg, M.A.Kaashoek, "Reduction of the abstract four block problem to a Nehari problem" is dedicated to the four-block extension problem. Here this problem in an abstract setting is reduced to a Nehari problem. The band method serves as a tool.

The sixth paper, A.B.Kuijper, "The state space method for integro-differential equations of Wiener-Hopf type with rational matrix symbols" is dedicated to the theory of Wiener-Hopf integro-differential systems of equations. In it are analyzed and explicitly solved such systems, where the symbol is a rational matrix function. The method is based on the realization of the symbol.

In the seventh paper, H.Widom, "Symbols and asymptotic expansions" is proposed a general principle which enables the computation of asymptotic expansions for the trace of functions of Wiener-Hopf or pseudodifferential operators. This principle is based on the use of the appropriate symbol in the continuous and discontinuous cases.

I.Gohberg

Tel-Aviv, June 1992

Operator Theory:
Advances and Applications, Vol. 58

UNCERTAINTY PRINCIPLES FOR TIME-FREQUENCY OPERATORS

John Benedetto*, Christopher Heil†, and David Walnut‡

Abstract: This paper explores some of the connections between classical Fourier analysis and time-frequency operators, as related to the role of the uncertainty principle in Gabor and wavelet basis expansions.

1. Introduction.

This paper will explore some of the connections between classical Fourier analysis and recent results in the area of time-frequency analysis, specifically Gabor and wavelet theory. Our starting point is the Poisson Summation Formula, which connects Fourier series on the circle with Fourier transforms on the line. This extraordinarily useful formula can be used, for example, to derive sampling formulas for band-limited functions and estimates for aliasing error when an arbitrary function is sampled. Following Section 1A, which contains a review of basic Fourier analysis, we present in Section 1B a proof of this sampling theorem and, as a corollary, the sampling theorem of Shannon and Whittaker. Next we consider the Gabor transform, which separates the time and frequency content of a function by using a windowed Fourier transform. The sampling theory of this transform can be cast in the language of decompositions of arbitrary functions in sets of basis-like functions and leads us to a discussion of frames and unconditional bases for Hilbert spaces. The wavelet transform also separates time and frequency content in a function

*Also Professor of Mathematics, University of Maryland, College Park; acknowledges partial support from NSF Grant DMS-9002420.
†Also Pure Mathematics Instructor, Massachusetts Institute of Technology; acknowledges partial support from NSF Grant DMS-9007212.
‡Also Assistant Professor of Mathematics, George Mason University, Fairfax, Virginia.

but substitutes scale for frequency and does not explicitly use a Fourier transform. The sampling theory here leads to the notion of multiresolution analysis and the construction of wavelet orthonormal bases. Sections 2A, B, and C contain the statements and proofs of theorems related to these two time-frequency operators.

Underlying our entire discussion is the classical uncertainty principle, which limits the degree to which a function can be localized both in time and frequency. The sampling and decomposition formulas outlined in the first two sections of the paper can be seen as the realization of an arbitrary function as combinations of functions well-localized in time and frequency. The uncertainty principle leads us to ask how well-localized those basis-like functions can be. For band-limited functions and the Fourier transform, the sparsest set of sampling which will permit exact reconstruction requires functions that are not well-localized in time, namely $\frac{\sin(\pi x)}{\pi x}$ and its translates. A denser sampling set leads to better localized basis functions. For the Gabor transform, the Balian–Low theorem tells us how well we can do for bases of Gabor functions, namely, that simultaneous quadratic localization is not possible in both time and frequency. In Section 3A we present an operator-theoretic proof of this theorem due to Battle. In this case, as with the previous one, allowing denser sets of sampling and overdetermined systems (frames) enables us to get better localization. For the wavelet transform, we present a generalization of a theorem due to Battle which states that any wavelet basis cannot have simultaneous exponential localization in time and frequency.

A. Basic Fourier analysis. Let $L^2[-T/2, T/2)$ denote the Hilbert space of T-periodic, square-integrable functions. Given $f \in L^2[-T/2, T/2)$, the Fourier coefficients of f are

$$c_n(f) = \frac{1}{\sqrt{T}} \int_{-T/2}^{T/2} f(t)\, e^{-2\pi i n t/T}\, dt.$$

Then we have

$$f(t) = \frac{1}{\sqrt{T}} \sum_n c_n(f)\, e^{2\pi i n t/T}, \tag{1.1}$$

where the sum converges in $L^2[-T/2, T/2)$. The Plancherel formula is

$$\int_{-T/2}^{T/2} |f(t)|^2\, dt = \sum_n |c_n(f)|^2,$$

and the Parseval formula is

$$\int_{-T/2}^{T/2} f(t)\,\overline{g(t)}\,dt \;=\; \sum_n c_n(f)\,\overline{c_n(g)},$$

for every $f,\, g \in L^2[-T/2, T/2)$.

Let $L^2(\mathbf{R})$ denote the Hilbert space of functions on the line which are square-integrable. Given $f \in L^2(\mathbf{R})$, the Fourier transform of f is

$$\hat{f}(\gamma) \;=\; \int f(t)\,e^{-2\pi i\gamma t}\,dt,$$

where the integral is defined as an L^2-limit in the usual way if it does not converge absolutely. Then we have

$$f(t) \;=\; \int \hat{f}(\gamma)\,e^{2\pi it\gamma}\,d\gamma,$$

where, as before, the integral is defined as an L^2-limit if it does not converge absolutely. The Plancherel formula is

$$\int |f(t)|^2\,dt \;=\; \int |\hat{f}(\gamma)|^2\,d\gamma$$

and the Parseval formula is

$$\int f(t)\,\overline{g(t)}\,dt \;=\; \int \hat{f}(\gamma)\,\overline{\hat{g}(\gamma)}\,d\gamma$$

for every $f,\, g \in L^2(\mathbf{R})$.

B. The Poisson Summation Formula. We define PW_Ω to be the space of functions $f \in L^2(\mathbf{R})$ such that $\mathrm{supp}(\hat{f}) \subset [-\Omega, \Omega]$. We call this the space of Ω-band-limited functions.

The Poisson Summation Formula (PSF) is:

THEOREM 1.1. *Let $f \in L^2(\mathbf{R})$ be such that*

$$|f(t)| \;\le\; C(1+|t|)^{-\alpha} \qquad \text{and} \qquad |\hat{f}(\gamma)| \;\le\; C(1+|\gamma|)^{-\alpha} \tag{1.2}$$

for some $C > 0$ and $\alpha > 1$. Then f and $\hat{f}$ are continuous functions, and, for any $P > 0$,

$$\sum_n f(t+nP) \;=\; \frac{1}{P}\sum_n \hat{f}(n/P)\,e^{2\pi int/P} \tag{1.3}$$

and

$$\sum_n \hat{f}(\gamma + nP) \;=\; \frac{1}{P}\sum_n f(n/P)\, e^{-2\pi i n\gamma/P}. \tag{1.4}$$

PROOF: By (1.2), the function $F(t) = \sum f(t+nP)$, the P-periodization of f, converges absolutely and uniformly on compact sets and is in $L^2[-P/2, P/2)$. Moreover,

$$\int_{-P/2}^{P/2} \sum_n f(t+nP)\, e^{-2\pi i n t/P}\, dt \;=\; \int f(t)\, e^{-2\pi i n t/P}\, dt \;=\; \hat{f}(n/P).$$

Therefore, by (1.1),

$$F(t) \;=\; \frac{1}{\sqrt{P}}\sum_n c_n(F)\, e^{2\pi i n t/P} \;=\; \frac{1}{P}\sum_n \hat{f}(n/P)\, e^{2\pi i n t/P},$$

so (1.3) holds. Equation (1.4) follows similarly. ∎

If $f \in L^2(\mathbf{R})$ is compactly supported then the P-periodization of f is an element of $L^2[-P/2, P/2)$. In this case, (1.3) holds as written with the sums on each side converging in $L^2[-P/2, P/2)$, and the assumption (1.2) is not required. The right side of (1.3) is the Fourier series of the function defined by the sum on the left. Similarly, if $f \in PW_\Omega$ then (1.4) holds as written with the sums converging in $L^2[-P/2, P/2)$.

We now state and prove as a corollary to the PSF the sampling formula of Shannon and Whittaker.

THEOREM 1.2. *Let $f \in PW_\Omega$. Then given $T \geq 2\Omega$,*

$$f(t) \;=\; \frac{1}{T}\sum_n f(n/T)\, \frac{\sin \pi(Tt-n)}{\pi(Tt-n)}.$$

Moreover,

$$\int |f(t)|^2\, dt \;=\; \frac{1}{T}\sum_n |f(n/T)|^2. \tag{1.5}$$

PROOF: By (1.3) and the remarks following Theorem 1.1,

$$\sum_n \hat{f}(\gamma + nT) \;=\; \frac{1}{T}\sum_n f(n/T)\, e^{-2\pi i n\gamma/T}.$$

We can write

$$\int_{-T/2}^{T/2} \sum_n \hat{f}(\gamma + nT)\, e^{2\pi i t\gamma}\, d\gamma \;=\; \frac{1}{T}\int_{-T/2}^{T/2} \sum_n f(n/T)\, e^{-2\pi i \gamma(n/T - t)}\, d\gamma. \tag{1.6}$$

Because the left hand side of (1.6) converges in $L^1[-T/2, T/2)$ we have

$$\int_{-T/2}^{T/2} \sum_n \hat{f}(\gamma + nT)\, e^{2\pi i t\gamma}\, d\gamma \;=\; \int_{-T/2}^{T/2} \hat{f}(\gamma)\, e^{2\pi i t\gamma}\, d\gamma \;=\; f(t).$$

Also, since the right hand side of (1.6) is the Fourier series of a function in $L^2[-T/2, T/2)$ we can integrate term-by-term, obtaining

$$\sum_n f(n/T)\, \frac{1}{T} \int_{-T/2}^{T/2} e^{2\pi i\gamma(t-n/T)}\, d\gamma \;=\; \sum_n f(n/T)\, \frac{\sin \pi(Tt-n)}{\pi(Tt-n)}.$$

Finally, (1.5) follows since $\{\frac{\sqrt{T}\sin \pi(Tt-n)}{\pi(Tt-n)}\}$ is an orthonormal set in $L^2[-T/2, T/2)$. ∎

We conclude this section with a variant of the sampling formula which is in keeping with the spirit of this paper. $\mathcal{S}(\mathbf{R})$ denotes the Schwartz class of infinitely differentiable, rapidly decreasing functions on the line.

THEOREM 1.3. *Let $f \in PW_\Omega$ and let $T > 2\Omega$ be given. Then there exists a function $s \in \mathcal{S}(\mathbf{R})$ such that*

$$f(t) \;=\; \frac{1}{T} \sum_n f(n/T)\, s(t - n/T).$$

PROOF: Let s be be any function such that $\hat{s} \in C^\infty(\mathbf{R})$, $\hat{s} = 1$ on $[-\Omega, \Omega]$, and $\hat{s} = 0$ outside $[-T/2, T/2]$. Then clearly $s \in \mathcal{S}(\mathbf{R})$ and $\hat{f}(\gamma)\,\hat{s}(\gamma) = \hat{f}(\gamma)$. The PSF implies that

$$\int \sum_n \hat{f}(\gamma + nT)\, \hat{s}(\gamma)\, e^{2\pi i t\gamma}\, d\gamma \;=\; \frac{1}{T} \int \sum_n f(n/T)\, e^{-2\pi i\gamma(n/T - t)}\, \hat{s}(\gamma)\, d\gamma. \tag{1.7}$$

Using an argument similar to one used in the proof of Theorem 1.2, the left hand side of (1.7) is

$$\int \hat{f}(\gamma)\, \hat{s}(\gamma)\, e^{2\pi i t\gamma}\, d\gamma \;=\; \int \hat{f}(\gamma)\, e^{2\pi i t\gamma}\, d\gamma \;=\; f(t),$$

and the right hand side of (1.7) is

$$\sum_n f(n/T)\, \frac{1}{T} \int \hat{s}(\gamma)\, e^{2\pi i\gamma(t-n/T)}\, d\gamma \;=\; \frac{1}{T} \sum_n f(n/T)\, s(t - n/T). \quad ∎$$

2. Sampling Results for Time-Frequency Transformations.

A. Time-frequency operators and frames. In this section we investigate some results concerning two time-frequency transformations, the continuous Gabor transform and the continuous wavelet transform. These operators are closely related to the Fourier transform and display simultaneously the time and frequency content of a signal.

Definition 2.1. Let $g \in L^2(\mathbf{R})$ be fixed. The *continuous Gabor transform* (*CGT*) with analyzing function g, denoted Ψ_g, is defined by

$$\Psi_g f(t,\omega) = \int_{-\infty}^{\infty} f(x)\, e^{-2\pi i \omega x}\, \overline{g(x-t)}\, dx,$$

where $f \in L^2(\mathbf{R})$.

In the above definition, t is the time variable and ω the frequency variable. The transform $\Psi_g f(t,\omega)$ is formed by shifting the window function g so that it is centered at t, then taking the Fourier transform. In this way, $\Psi_g f(t,\omega)$ displays the frequency content of f near time t.

The definition of the continuous wavelet transform uses the same idea of a sliding window but replaces the frequency variable by a scale variable.

Definition 2.2. Let $\psi \in L^2(\mathbf{R})$ be fixed. The *continuous wavelet transform* (*CWT*) with analyzing function ψ, denoted Φ_ψ, is defined by

$$\Phi_\psi f(a,b) = \frac{1}{a} \int_{-\infty}^{\infty} f(x)\, \overline{\psi\Big(\frac{x-b}{a}\Big)}\, dx,$$

where $f \in L^2(\mathbf{R})$.

Note that Φ_ψ is a time-frequency localization operator where a is the scale (frequency) variable and b the time variable.

We now present some elementary properties of the CGT and CWT. Proofs of these results may be found in numerous places, e.g., [**D1; HW**]. We point out here that the results are immediate consequences of the Plancherel and Parseval formulas for the Fourier transform.

THEOREM 2.3. *Let $g \in L^2(\mathbf{R})$. Then for all $f \in L^2(\mathbf{R})$,*

$$\int_{-\infty}^{\infty}\int_{-\infty}^{\infty} |\Psi_g f(t,\omega)|^2\, dt\, d\omega = \int_{-\infty}^{\infty} |g(x)|^2\, dx \int_{-\infty}^{\infty} |f(x)|^2\, dx.$$

THEOREM 2.4. *Let $\psi \in L^2(\mathbf{R})$ and suppose that*

$$\int_{-\infty}^{\infty} \frac{|\hat{\psi}(\gamma)|^2}{|\gamma|}\, d\gamma < \infty.$$

Then

$$\int_0^{\infty}\int_{-\infty}^{\infty} |\Phi_\psi f(a,b)|^2\, db\, \frac{da}{a}$$
$$= \int_0^{\infty} |\hat{f}(\omega)|^2\, d\omega \int_0^{\infty} \frac{|\hat{\psi}(\gamma)|^2}{|\gamma|}\, d\gamma + \int_{-\infty}^{0} |\hat{f}(\omega)|^2\, d\omega \int_{-\infty}^{0} \frac{|\hat{\psi}(\gamma)|^2}{|\gamma|}\, d\gamma.$$

The Poisson Summation Formula can be seen as a sampling result which characterizes the recovery of a function from regular samples of its Fourier transform. In some cases the recovery is exact, for example when f is supported in the interval $[0,1]$. Such a sampling result can be cast in the language of Fourier series, i.e., in the representation of a function as a superposition of complex exponentials. In the case of the time-frequency transformations considered here, there is a well-developed theory of the recovery of a function from samples of its CGT or CWT. Here recovery is exact in all of $L^2(\mathbf{R})$, and, as in the Fourier series case, sampling results can be cast in the language of the representation of a function as a superposition of a fixed collection of basic functions. The most convenient setting for such representations is that of a *frame* in a Hilbert space, first described by Duffin and Schaeffer [**DS**] in relation to non-harmonic Fourier series.

DEFINITION 2.5. Let H be a separable Hilbert space. A collection $\{x_n\} \subset H$ is a *frame* if there exist constants A, $B > 0$ such that for all $x \in H$,

$$A\,\|x\|^2 \le \sum_n |\langle x, x_n\rangle|^2 \le B\,\|x\|^2.$$

A frame is said to be *tight* if $A = B$ and *exact* if $\{x_n\}$ ceases to be a frame upon the removal of any element.

We say $\{x_n\}$ is a *Riesz basis* if there exist constants A, $B > 0$ such that $A \le \|x_n\| \le B$ for all n and if every $x \in H$ can be written $x = \sum c_n(x)\, x_n$ for a unique choice of scalars $\{c_n(x)\}$ and this sum converges unconditionally (every rearrangement of the sum also converges, and to the same value).

A frame is a generalization of an orthonormal basis or a Riesz basis. With a frame, one has representations of elements in the Hilbert space in terms of the frame, and one can recover elements in a constructive way from the frame coefficients $\{\langle x, x_n\rangle\}$. The difference is that frames may be overcomplete in the sense that each element of a frame is in the closed linear span of the remaining elements. The following theorem collects some standard results about frames, proofs of which can be found in, e.g., [**DS**; **HW**; **D1**].

THEOREM 2.6. *Let H be a separable Hilbert space and $\{x_n\}$ a collection of vectors in H. Then*

(a) *$\{x_n\}$ is a frame for H if and only if the operator S defined by $Sx = \sum\langle x, x_n\rangle\, x_n$ converges in H for each x and is a topological isomorphism of H onto itself. In this case, we have the identities $x = \sum\langle x, x_n\rangle\, S^{-1}x_n = \sum\langle x, S^{-1}x_n\rangle\, x_n$, and $\{S^{-1}x_n\}$ is also a frame for H, called the **dual frame** of $\{x_n\}$.*

(b) *$\{x_n\}$ is an exact frame if and only if $\{x_n\}$ is a Riesz basis for H.*

(c) *If $\{x_n\}$ is such that $\|x_n\| = 1$ for all n, and if $\{x_n\}$ forms a tight frame with $A = B = 1$, then $\{x_n\}$ is an orthonormal basis for H.*

PROOF: For a proof of (a), see, e.g., [**HW**, Theorem 2.1.3]. For (b), see [**HW**, Theorem 2.2.2]. To see (c), observe that for n fixed,

$$\|x_n\|^2 = \sum_k |\langle x_n, x_k\rangle|^2 = \|x_n\|^4 + \sum_{k\neq n} |\langle x_n, x_k\rangle|^2.$$

Since $\|x_n\| = 1$, $\langle x_n, x_k\rangle = 0$ for all $n \neq k$. Thus $\{x_n\}$ is an orthonormal system. Because of (a), the span of $\{x_n\}$ is dense in H, i.e., $\{x_n\}$ is complete in H. A complete orthonormal sequence is an orthonormal basis. ∎

B. Frames and sampling for the CGT. In this section we investigate the connections between sampling results for the CGT and the existence of frames in the Hilbert space $L^2(\mathbf{R})$.

DEFINITION 2.7. Let $a, b > 0$ and $g \in L^2(\mathbf{R})$ be fixed. Then if the collection $\{e^{2\pi imbx}g(x-na)\}$ is a frame for $L^2(\mathbf{R})$, it is called a *Gabor frame*. We will denote the elements of such a frame by $\{E_{mb}T_{na}g\}$, where

$$T_xg(t) = g(t-x) \quad \text{and} \quad E_yg(t) = e^{2\pi iyt}g(t).$$

In his fundamental paper [**G**], Gabor investigated the case when $a = b = 1$ and $g(x) = e^{-\pi x^2}$ (cf., Section 3A). In this case, the collection $\{E_mT_ng\}$ is complete in $L^2(\mathbf{R})$ but does not form a frame. The idea of applying the notion of a frame to sets of the form $\{E_{mb}T_{na}g\}$ was first described in [**DGM**].

The following result can be viewed as a sampling theorem for the CGT. Its proof is immediate from the definition of a Gabor frame.

THEOREM 2.8. *The collection $\{E_{mb}T_{na}g\}$ is a Gabor frame if and only if there exist constants $0 < A < B$ such that for all $f \in L^2(\mathbf{R})$,*

$$A\,\|f\|_2^2 \le \sum_n\sum_m |\Psi_gf(na,mb)|^2 \le B\,\|f\|_2^2.$$

We now introduce the Zak transform, which we will use to partially characterize Gabor frames (see [**AT; D; HW; J**]). We let Q denote the unit square, i.e., $Q = [0,1)^2$.

DEFINITION 2.9. Given $a > 0$, the **Zak transform**, or **Weil-Brezin map**, denoted Z_a, is the unitary mapping from $L^2(\mathbf{R})$ onto $L^2(Q)$ given by

$$Z_af(t,\omega) = a^{1/2}\sum_k e^{2\pi ik\omega}\,f(a(t-k)).$$

Z_af satisfies the following *quasi-periodicity* relations:

$$Z_af(t+1,\omega) = e^{2\pi i\omega}Z_af(t,\omega)$$
$$Z_af(t,\omega+1) = Z_af(t,\omega).$$

LEMMA 2.10. *Suppose that $ab = 1/N$ for some integer $N \ge 1$. Then*

$$Z_a(E_{mb}T_{na}g)(t,\omega) = e^{2\pi imn/N}\,e^{2\pi imt/N}\,e^{-2\pi in\omega}\,Z_ag(t,\omega - m/N).$$

THEOREM 2.11. *Let $g \in L^2(\mathbf{R})$, and suppose that $ab = 1/N$ for some $a,\ b > 0$, $N \in \mathbf{Z}$. Then in order that $\{E_{mb}T_{na}g\}$ be a Gabor frame, it is necessary and sufficient that there exist constants $A,\ B > 0$ such that*

$$A \le \sum_{j=0}^{N-1} |Z_ag(t,\omega - j/N)|^2 \le B. \tag{2.1}$$

PROOF: Given $f \in L^2(\mathbf{R})$, we have by the unitarity of Z_a, Lemma 2.10, and Plancherel's formula for Fourier series,

$$\begin{aligned}
&\sum_n \sum_m |\langle f, E_{mb}T_{na}g\rangle|^2 \\
&= \sum_n \sum_m \left| \int_0^1 \int_0^1 Z_a f(t,\omega)\, \overline{Z_a g(t,\omega - m/N)}\, e^{-2\pi i m t/N}\, e^{2\pi i n \omega}\, dt\, d\omega \right|^2 \\
&= \sum_{j=0}^{N-1} \sum_n \sum_k \left| \int_0^1 \int_0^1 Z_a f(t,\omega)\, \overline{Z_a g(t,\omega - j/N)}\, e^{-2\pi i j t/N}\, e^{-2\pi i k t}\, e^{2\pi i n \omega}\, dt\, d\omega \right|^2 \\
&= \int_0^1 \int_0^1 |Z_a f(t,\omega)|^2 \sum_{j=0}^{N-1} |Z_a g(t,\omega - j/N)|\, dt\, d\omega.
\end{aligned}$$

Therefore, if (2.1) holds then

$$A\, \|Z_a f\|^2_{L^2(Q)} \;\le\; \int_0^1 \int_0^1 |Z_a f(t,\omega)|^2 \sum_{j=0}^{N-1} |Z_a g(t,\omega - j/N)|\, dt\, d\omega \;\le\; B\, \|Z_a f\|^2_{L^2(Q)},$$

which by the unitarity of Z_a is

$$A\|f\|_2^2 \;\le\; \sum_n \sum_m |\langle f, E_{mb}T_{na}g\rangle|^2 \;\le\; B\|f\|_2^2.$$

Conversely, suppose (2.1) does not hold, say that

$$\operatorname*{ess\,inf}_{(t,\omega)\in Q} \sum_{j=0}^{N-1} |Z_a g(t,\omega - j/N)|^2 \;=\; 0.$$

Then given $\epsilon > 0$, there is a set $E \subset Q$ of positive measure such that

$$\sum_{j=0}^{N-1} |Z_a g(t,\omega - j/N)|^2 \;<\; \epsilon$$

for $(t,\omega) \in E$. Let $f \in L^2(\mathbf{R})$ be such that

$$Z_a f(t,\omega) \;=\; \frac{1}{|E|^{1/2}}\, \chi_E(t,\omega)$$

for $(t,\omega) \in Q$. Then $\|f\|_2 = \|Z_a f\|_{L^2(Q)} = 1$, but

$$\int_0^1 \int_0^1 |Z_a f(t,\omega)|^2 \sum_{j=0}^{N-1} |Z_a g(t,\omega - j/N)|\, dt\, d\omega \;=\; \sum_n \sum_m |\langle f, E_{mb}T_{na}g\rangle|^2 \;<\; \epsilon.$$

That is, $\{E_{mb}T_{na}g\}$ fails to have a lower frame bound. A similar argument shows that $\{E_{mb}T_{na}g\}$ fails to have an upper frame bound if

$$\operatorname*{ess\,sup}_{(t,\omega)\in Q} \sum_{j=0}^{N-1} |Z_a g(t,\omega - j/N)|^2 \;=\; +\infty. \quad \blacksquare$$

C. Frames and sampling for the CWT. In this section we investigate the connections between sampling results for the CWT and the existence of orthonormal bases of wavelets in $L^2(\mathbf{R})$.

DEFINITION 2.12. Let $\psi \in L^2(\mathbf{R})$ be fixed. Then if the collection $\{2^{j/2}\,\psi(2^j x - k)\}$ is an orthonormal basis for $L^2(\mathbf{R})$, it is called a *wavelet orthonormal basis.*

For simplicity, we will write $\psi_{jk}(x) = 2^{j/2}\,\psi(2^j x - k)$. The existence of wavelet orthonormal bases has been known since Haar, who showed that if $\psi = \chi_{[0,1/2)} - \chi_{[1/2,1)}$ then $\{\psi_{jk}\}$ is a wavelet orthonormal basis, known as the *Haar system.* More recently, Meyer has shown that there exists ψ, infinitely differentiable and rapidly decreasing such that $\{\psi_{jk}\}$ is a wavelet orthonormal basis (see Section 3B) and Daubechies has shown that for any $r > 0$, there exist ψ, compactly supported and r–times continuously differentiable such that $\{\psi_{jk}\}$ is a wavelet orthonormal basis. Wavelet *frames*, especially those which are exact, are discussed in Section 3B.

The following result can be viewed as a sampling theorem for the CWT. Its proof is immediate from the definition of a wavelet orthonormal basis.

THEOREM 2.13. *If $\psi \in L^2(\mathbf{R})$ is such that $\|\psi\|_2 = 1$, then $\{\psi_{jk}\}$ is a wavelet orthonormal basis if and only if*

$$\|f\|_2^2 = \sum_j \sum_k |\Phi_\psi f(2^{-j}, 2^{-j}k)|^2.$$

In order to partially characterize wavelet orthonormal bases, we introduce the notion of a multiresolution analysis due to S. Mallat (see [**M; Ma; D2**]).

DEFINITION 2.14. A *multiresolution analysis* (MRA) of $L^2(\mathbf{R})$ is a collection of closed subspaces $\{V_j\}_{j\in\mathbf{Z}}$ such that the following conditions hold.

(a) $V_j \subset V_{j+1}$ for all j.

(b) $\cup V_j$ is dense in $L^2(\mathbf{R})$.

(c) $\cap V_j = \{0\}$.

(d) $f(x) \in V_j$ if and only if $f(2x) \in V_{j+1}$ for all j.

(e) There exists a function $\varphi \in L^2(\mathbf{R})$ such that the collection $\{\varphi(x-k)\}_{k\in\mathbf{Z}}$ is an orthonormal basis for V_0.

Given a MRA, one can construct a wavelet orthonormal basis.

LEMMA 2.15. *Let $g \in L^2(\mathbf{R})$. Then $\{g(x-k)\}_{k\in\mathbf{Z}}$ is an orthonormal set if and only if $\sum_{n\in\mathbf{Z}} |\hat{g}(\gamma+n)|^2 \equiv 1$.*

PROOF: $\{g(x-k)\}_{k\in\mathbf{Z}}$ is an orthonormal set if and only if $g(x)$ is orthogonal to $g(x-k)$ for all $k \neq 0$ and $\|g\|_2 = 1$. That is, if and only if

$$\int g(x)\overline{g(x-k)}\,dx \;=\; \int |\hat{g}(\gamma)|^2\, e^{-2\pi i k\gamma}\,d\gamma \;=\; \int_0^1 \sum_{n\in\mathbf{Z}} |\hat{g}(\gamma+n)|^2\, e^{-2\pi i k\gamma}\,d\gamma \;=\; \delta_k.$$

This is true if and only if $\sum_{n\in\mathbf{Z}} |\hat{g}(\gamma+n)|^2 \equiv 1$. ∎

The following result is due to Y. Meyer.

THEOREM 2.16. *Let $\{V_j\}$ be a MRA. Then there exists $\psi \in L^2(\mathbf{R})$ such that $\{\psi_{jk}\}$ is a wavelet orthonormal basis.*

PROOF: Since $V_j \subset V_{j+1}$, we may define W_j as the orthogonal complement of V_j in V_{j+1}, so that $V_{j+1} = V_j \oplus W_j$ for all j. Then $\{W_j\}$ is a collection of mutually orthogonal closed subspaces of $L^2(\mathbf{R})$. Let P_j and Q_j be the orthogonal projectors onto V_j and W_j, respectively. Then $P_{j+1} = P_j + Q_j$ for all j.

Now suppose that $\psi \in L^2(\mathbf{R})$ is given so that $\{\psi_{0k}\}$ is an orthonormal basis for W_0. Then we claim that $\{\psi_{jk}\}$ is a wavelet orthonormal basis for $L^2(\mathbf{R})$. To see this, note that $f(x) \in W_0$ if and only if $f(2^j x) \in W_j$. Therefore for each fixed j, $\{\psi_{jk}\}_{k\in\mathbf{Z}}$ is an orthonormal basis for W_j. In particular, $\{\psi_{jk}\}_{j,k\in\mathbf{Z}}$ is an orthonormal system. To see that $\{\psi_{jk}\}$ is complete, note that for each $f \in L^2(\mathbf{R})$, $P_j f \to f$ and $P_{-j} f \to 0$ as $j \to \infty$. Thus, $\|f - (P_j - P_{-j})f\|_2 \to 0$ as $j \to \infty$. However,

$$P_j - P_{-j} \;=\; \sum_{k=-j}^{j-1} (P_{k+1} - P_k) \;=\; \sum_{k=-j}^{j-1} Q_k.$$

Since $Q_j f$ lies within the closed span of $\{\psi_{jk}\}_{k\in\mathbf{Z}}$ for each j, we conclude that the span of $\{\psi_{jk}\}_{j,k\in\mathbf{Z}}$ is dense in all of $L^2(\mathbf{R})$ and that $\{\psi_{jk}\}_{j,k\in\mathbf{Z}}$ is a wavelet orthonormal basis.

In order to construct such a ψ, note that since $\varphi \in V_0 \subset V_1$ and since $\{\varphi_{1k}\}_{k\in\mathbf{Z}}$ is an orthonormal basis for V_1, φ satisfies the *dilation equation*

$$\varphi(x) \;=\; \sum_k h_k\, \varphi(2x-k),$$

where $h_k = 2^{1/2}\,\langle \varphi, \varphi_{1k}\rangle$. Now define

$$\psi(x) \;=\; \sum_k (-1)^k\, h_{1-k}\, \varphi(2x-k).$$

We claim that $\{\psi_{0k}\}$ is an orthonormal basis for W_0. In order to see this, define

$$m_0(\gamma) \;=\; \tfrac{1}{2}\sum_k h_k\, e^{-2\pi i k\gamma} \qquad \text{and} \qquad m_1(\gamma) \;=\; \tfrac{1}{2}\sum_k (-1)^k\, h_{1-k}\, e^{-2\pi i k\gamma}. \tag{2.2}$$

Then

$$\begin{aligned} \hat{\varphi}(\gamma) &= m_0(\gamma/2)\,\hat{\varphi}(\gamma/2), \\ \hat{\psi}(\gamma) &= m_1(\gamma/2)\,\hat{\varphi}(\gamma/2), \\ m_1(\gamma) &= e^{2\pi i(\gamma+1/2)}\,\overline{m_0(\gamma+1/2)}. \end{aligned}$$

By Lemma 2.15, since $\{\varphi_{0n}\}$ is an orthonormal set,

$$\begin{aligned} &\sum_n |\hat{\varphi}(\gamma+n)|^2 \\ &= \sum_n |m_0(\gamma/2+n/2)|^2\, |\hat{\varphi}(\gamma/2+n/2)|^2 \\ &= |m_0(\gamma/2)|^2 \sum_{n \text{ even}} |\hat{\varphi}(\gamma/2+n/2)|^2 \;+\; |m_0(\gamma/2+1/2)|^2 \sum_{n \text{ odd}} |\hat{\varphi}(\gamma/2+n/2)|^2 \\ &= |m_0(\gamma/2)|^2 \sum_j |\hat{\varphi}(\gamma/2+j)|^2 \;+\; |m_0(\gamma/2+1/2)|^2 \sum_j |\hat{\varphi}(\gamma/2+j+1/2)|^2 \\ &= |m_0(\gamma/2)|^2 \;+\; |m_0(\gamma/2+1/2)|^2. \end{aligned}$$

Therefore, by Lemma 2.15,

$$|m_0(\gamma/2)|^2 \;+\; |m_0(\gamma/2+1/2)|^2 \;\equiv\; 1.$$

And, since $|m_1(\gamma)| = |m_0(\gamma+1/2)|$,

$$|m_1(\gamma/2)|^2 \;+\; |m_1(\gamma/2+1/2)|^2 \;\equiv\; 1.$$

Then, as above,

$$\sum_n |\hat{\psi}(\gamma+n)|^2 \;=\; |m_1(\gamma/2)|^2 \;+\; |m_1(\gamma/2+1/2)|^2 \;\equiv\; 1,$$

so that $\{\psi_{0k}\}$ is an orthonormal set by Lemma 2.15.

It remains to show that $W_0 = \overline{\text{span}}\{\psi_{0k}\}$. Now, for any $f \in L^2(\mathbf{R})$,

$$P_0 f = \sum_k c_k \varphi_{0k}$$

where $c_k = \langle f, \varphi_{0k} \rangle$, and

$$(P_0 f)\hat{}(\gamma) = c(\gamma)\hat{\varphi}(\gamma) = c(\gamma) m_0(\gamma/2)\hat{\varphi}(\gamma/2), \tag{2.3}$$

where $c(\gamma) = \sum c_k e^{-2\pi i k\gamma} \in L^2[0,1]$. Similarly,

$$(P_1 f)\hat{}(\gamma) = a(\gamma/2)\hat{\varphi}(\gamma/2), \tag{2.4}$$

where $a(\gamma) = 2^{-1/2} \sum a_k e^{-2\pi i k\gamma} \in L^2[0,1]$ and $a_k = \langle f, \varphi_{1k} \rangle$.

Consider the system of equations

$$\begin{pmatrix} m_0(\gamma) & m_1(\gamma) \\ m_0(\gamma+1/2) & m_1(\gamma+1/2) \end{pmatrix} \begin{pmatrix} c(\gamma) \\ d(\gamma) \end{pmatrix} = \begin{pmatrix} a(\gamma) \\ a(\gamma+1/2) \end{pmatrix}, \tag{2.5}$$

where $a(\gamma) \in L^2[0,1]$ is given. The determinant of the matrix on the left reduces to

$$e^{2\pi i(\gamma+1/2)} \left(|m_0(\gamma)|^2 + |m_0(\gamma+1/2)|^2\right) = e^{2\pi i(\gamma+1/2)},$$

which is non-zero for all γ. The solution of the system is therefore

$$c(\gamma) = e^{-2\pi i(\gamma+1/2)} \left(m_1(\gamma+1/2)a(\gamma) - m_1(\gamma)a(\gamma+1/2)\right),$$

$$d(\gamma) = e^{-2\pi i(\gamma+1/2)} \left(m_0(\gamma)a(\gamma+1/2) - m_0(\gamma+1/2)a(\gamma)\right).$$

Since m_0 and m_1 are bounded, c, $d \in L^2[0,1]$.

Now, for any $f \in L^2(\mathbf{R})$, $Q_0 f = P_1 f - P_0 f$, and so by (2.3) and (2.4),

$$(Q_0 f)\hat{}(\gamma) = \left(a(\gamma/2) - c(\gamma) m_0(\gamma/2)\right) \hat{\varphi}(\gamma/2).$$

However, by (2.5), $a(\gamma/2) - c(\gamma) m_0(\gamma/2) = d(\gamma) m_1(\gamma/2)$, so that

$$(Q_0 f)\hat{}(\gamma) = d(\gamma) m_1(\gamma/2)\hat{\varphi}(\gamma/2) = d(\gamma)\hat{\psi}(\gamma)$$

for some $d(\gamma) = \sum d_k e^{-2\pi i k\gamma} \in L^2[0,1]$. Thus, $Q_0 f = \sum d_k \psi_{0k}$, which implies that the span of $\{\psi_{0k}\}$ is dense in W_0, and hence that $\{\psi_{jk}\}$ is a wavelet orthonormal basis for $L^2(\mathbf{R})$. ∎

In some cases it is possible to define a MRA starting with the auxiliary function m_0 defined by (2.2). The following result of Cohen [C], which we state without proof, characterizes when this is possible.

THEOREM 2.17. *Let h_k be a real-valued sequence and suppose that*

(a) $\sum |h_k|\, k^\epsilon < \infty$ *for some* $\epsilon > 0$,

(b) $m_0(0) = 1$,

(c) $|m_0(\gamma)|^2 + |m_0(\gamma + 1/2)|^2 = 1$ *for all* γ.

Let φ be defined by

$$\hat{\varphi}(\gamma) = \prod_{j=1}^{\infty} m_0(\gamma/2^j).$$

Then $\{\varphi(x-k)\}$ is an orthonormal set if and only if there exists a compact set K such that

(d) *K is a finite union of closed intervals of total measure 1.*

(e) *for every $x \in [-1/2, 1/2]$ there is an integer k such that $x + k \in K$,*

(f) *K contains a neighborhood of zero,*

(g) *there exists $\epsilon > 0$ such that $|m_0(2^{-k}\gamma)| \geq \epsilon$ for all $k \geq 1$ and $\gamma \in K$.*

In this case, taking $V_0 = \overline{\operatorname{span}}\{\varphi(x-k)\}$ and defining $V_j = \{f(2^{-j}x) : f \in V_0\}$, the $\{V_j\}$ are a MRA, and therefore a wavelet orthonormal basis exists for $L^2(\mathbf{R})$.

In the special case obtained by putting $K = [-1/2, 1/2]$, conditions (d)–(g) are satisfied if and only if $|m_0(\gamma)| > 0$ for $|\gamma| < 1/4$. This fact was proved by Mallat [**Ma**].

3. UNCERTAINTY PRINCIPLES FOR EXACT GABOR AND WAVELET FRAMES.

The previous sections described several time-frequency analysis techniques, including Gabor and wavelet frames. In applications, exact frames (especially those which are actually orthonormal bases) are often appealing, as they can often be implemented with fast discrete algorithms. We focus on such exact systems in this section, establishing some limitations on the joint time-frequency concentration of the elements of Gabor and wavelet systems which form exact frames. These results may thus be considered "uncertainty principles" for such nonredundant systems.

First, however, we review some general results on exact frames and the classical uncertainty principle in Hilbert spaces. By Theorem 2.6(a), any exact frame $\{x_n\}$ for a Hilbert space H is a *Riesz basis* for H. Moreover, $c_n(x) = \langle x, \tilde{x}_n \rangle$ where $\{\tilde{x}_n\}$ is the dual frame to $\{x_n\}$, i.e., $\tilde{x}_n = S^{-1}x_n$. Therefore, $\{x_n\}$ and $\{\tilde{x}_n\}$ are *biorthonormal*, i.e., $\langle x_m, \tilde{x}_n\rangle = \delta_{mn}$.

The following inequality is the *classical uncertainty principle.* Its proof, which is quite elementary, was recorded by N. Wiener and H. Weyl.

THEOREM 3.1. *If $f \in L^2(\mathbf{R})$ then $\|f\|_2^2 \leq 4\pi \, \|tf(t)\|_2 \, \|\gamma\hat{f}(\gamma)\|_2$.*

Given a Hilbert space H and given (not necessarily continuous) operators A, B mapping domains $D(A)$, $D(B) \subset H$ into H, respectively, define the *commutator* of A, B to be the operator $[A,B] = AB - BA$. If A is self-adjoint then the *expectation* of A at $f \in D(A)$ is $E_f(A) = \langle Af, f\rangle$, and the *variance* of A at $f \in D(A^2)$ is $\sigma_f^2(A) = E_f(A^2) - E_f(A)^2$. An uncertainty principle inequality can now be formulated on H. Its statement (Theorem 3.2) and simple computational proof are part of the folklore in the Hilbert space community.

THEOREM 3.2. *Given self-adjoint operators A, B on a Hilbert space H. If $f \in D(A^2) \cap D(B^2) \cap D(i[A,B])$ and $\|f\| = 1$ then $E_f(i[A,B])^2 \leq 4\,\sigma_f^2(A)\,\sigma_f^2(B)$.*

The classical uncertainty principle (Theorem 3.1) follows as a corollary. Define the *position operator* P (operating on functions f) by

$$Pf(t) \; = \; t\,f(t),$$

and the *momentum operator* M by

$$Mf \; = \; (P\hat{f})^{\vee} \; = \; (\gamma\,\hat{f}(\gamma))^{\vee}.$$

The domains of P, M include the Schwartz class $\mathcal{S}(\mathbf{R}) \subset L^2(\mathbf{R})$, and both P, M are self-adjoint. If $f \in \mathcal{S}(\mathbf{R})$ then

$$\begin{aligned} (f')^{\wedge} \; &= \; 2\pi i\, P\hat{f}, \\ f' \; &= \; 2\pi i\, Mf, \\ [P,M]f \; &= \; -\tfrac{1}{2\pi i} f. \end{aligned} \tag{3.1}$$

From Theorem 3.2 we therefore have

$$E_f(-\tfrac{1}{2\pi}I)^2 \; \leq \; \sigma_f^2(P)\,\sigma_f^2(M),$$

where I is the identity operator on $\mathcal{S}(\mathbf{R})$. However, $E_f(I) = \|f\|_2^2$, $\sigma_f^2(P) = \|Pf\|_2^2 - \langle Pf, f\rangle^2 \leq \|Pf\|_2^2$, and $\sigma_f^2(M) = \|Mf\|_2^2 - \langle Mf, f\rangle^2 \leq \|Mf\|_2^2$, from which Theorem 3.1 follows immediately for $f \in \mathcal{S}(\mathbf{R})$. A standard closure argument extends the inequality to all $f \in L^2(\mathbf{R})$.

A. An uncertainty principle for exact Gabor frames. If $ab = 1$ and $\{E_{mb}T_{na}g\}$ is a Gabor frame for $L^2(\mathbf{R})$ then it must be exact. Conversely, if $\{E_{mb}T_{na}g\}$ is an exact Gabor frame then necessarily $ab = 1$. By dilating g if necessary we therefore assume $a = b = 1$ in this section without loss of generality, and for simplicity write $g_{mn}(x) = E_mT_ng(x) = e^{2\pi imx}\, g(x-n)$, and let Z denote the Zak transform $Z = Z_a = Z_1$. Note that the dual frame to $\{g_{mn}\}$ is another exact Gabor frame $\{\tilde{g}_{mn}\}$, where $\tilde{g} = S^{-1}g$ is a uniquely determined function in $L^2(\mathbf{R})$.

The *Balian-Low Theorem* (BLT) is an uncertainty principle-like result for exact Gabor frames. It imposes severe restictions on the time-frequency localization of any function g which generates an exact Gabor frame:

THEOREM 3.3. *Given $g \in L^2(\mathbf{R})$. If $\{g_{mn}\}$ is an exact Gabor frame then*

$$\|t\, g(t)\|_2\ \|\gamma\, \hat{g}(\gamma)\|_2 \;=\; +\infty.$$

In particular, note that if g is the Gaussian function $g(x) = e^{-\pi x^2}$ then $\{g_{mn}\}$ is not an exact Gabor frame.

The BLT was first stated by Balian [**B**], and later (independently) by Low [**L**], for the special case of Gabor systems which are orthonormal bases. Their proofs contained a technical gap which was filled by Coifman, Semmes, and Daubechies [**D1**]; this group also extended the result to all exact Gabor frames. The resulting proof required the use of the Zak transform and, implicitly, both classical and distributional differentiation. Explicit justification for this proof, along with a non-distributional version, was presented in [**BHW**]. As all versions of this proof of the BLT are quite technical, we shall instead discuss a proof inspired by an elegant, natural, and elementary argument due to Battle [**Bat**], originally stated for the case of orthonormal bases only, extended formally to exact frames by Daubechies and Janssen [**DJ**], and justified in [**BHW**]. This argument depends critically on (3.1), from which the classical uncertainty principle can also be derived. Precisely,

(3.1) is used in the following form:

LEMMA 3.4. *Given f, $g \in L^2(\mathbf{R})$. If Pf, Pg, $P\hat{f}$, $P\hat{g} \in L^2(\mathbf{R})$ then*

$$\langle Pf, Mg\rangle - \langle Mf, Pg\rangle = \tfrac{1}{2\pi i}\,\langle f, g\rangle. \tag{3.2}$$

Note that (3.2) follows immediately from (3.1) for f, $g \in \mathcal{S}(\mathbf{R})$, and can be extended to all f, $g \in L^2(\mathbf{R})$ satisfying the hypotheses of Lemma 3.4 by a closure argument.

Lemma 3.4 suffices to prove a weak version of the BLT:

THEOREM 3.5. *Given $g \in L^2(\mathbf{R})$. If $\{g_{mn}\}$ is an exact Gabor frame then*

$$\|t\,g(t)\|_2\ \|\gamma\,\hat{g}(\gamma)\|_2\ \|t\,\tilde{g}(t)\|_2\ \|\gamma\,\hat{\tilde{g}}(\gamma)\|_2 = +\infty.$$

PROOF: Assume Pg, $P\hat{g}$, $P\tilde{g}$, $P\hat{\tilde{g}} \in L^2(\mathbf{R})$. An easy computation yields the formulas

$$\begin{aligned} [E_mT_n, P] &= -n\,E_mT_n, \\ [E_mT_n, M] &= -m\,E_mT_n. \end{aligned}$$

Since $\{g_{mn}\}$ and $\{\tilde{g}_{mn}\}$ are biorthonormal, and recalling that $g = g_{00}$ and $\tilde{g} = \tilde{g}_{00}$, it follows that

$$\begin{aligned} \langle g_{mn}, P\tilde{g}\rangle &= \langle E_mT_ng, P\tilde{g}\rangle = \langle E_mT_nPg, \tilde{g}\rangle = \langle (Pg)_{mn}, \tilde{g}\rangle, \\ \langle Mg, \tilde{g}_{mn}\rangle &= \langle Mg, E_mT_n\tilde{g}\rangle = \langle g, E_mT_nM\tilde{g}\rangle = \langle g, (M\tilde{g})_{mn}\rangle. \end{aligned}$$

Since $e^{2\pi imn} = 1$ for all m, n, we therefore have

$$\begin{aligned} \langle Pg, M\tilde{g}\rangle &= \sum \langle Pg, \tilde{g}_{mn}\rangle\,\langle g_{mn}, M\tilde{g}\rangle \\ &= \sum \langle (Pg)_{mn}, \tilde{g}\rangle\,\langle g, (M\tilde{g})_{mn}\rangle \\ &= \sum \langle g_{mn}, P\tilde{g}\rangle\,\langle Mg, \tilde{g}_{mn}\rangle \\ &= \langle Mg, P\tilde{g}\rangle. \end{aligned}$$

Therefore, $\langle g, \tilde{g}\rangle = 0$ by (3.2). However, we also have $\langle g, \tilde{g}\rangle = 1$ by biorthonormality, a contradiction. ∎

Note that Theorems 3.3 and 3.5 are equivalent if $\{g_{mn}\}$ is an orthonormal basis, for then $g = \tilde{g}$. We demonstrate now that Theorems 3.3 and 3.5 are equivalent in general, for which it suffices to prove:

PROPOSITION 3.6. *Given $g \in L^2(\mathbf{R})$. If $\{g_{mn}\}$ is an exact Gabor frame then*

$$Pg \in L^2(\mathbf{R}) \Leftrightarrow P\tilde{g} \in L^2(\mathbf{R}) \tag{3.3}$$

and

$$P\hat{g} \in L^2(\mathbf{R}) \Leftrightarrow P\hat{\tilde{g}} \in L^2(\mathbf{R}). \tag{3.4}$$

PROOF: We give only a formal argument, due to Daubechies and Janssen [**DJ**] for (3.3); the proof of (3.4) is entirely symmetrical.

Given $f \in L^2(\mathbf{R})$, we formally compute

$$\begin{aligned} ZPf(t,\omega) &= t \sum f(t+k)\, e^{2\pi i k\omega} + \sum f(t+k)\, k\, e^{2\pi i k\omega} \\ &= t\, Zf(t,\omega) + \tfrac{1}{2\pi i}\, \partial_2 Zf(t,\omega). \end{aligned} \tag{3.5}$$

Note that from Lemma 2.10, $Zg_{mn} = E_{(m,n)}Zg$ and $Z\tilde{g}_{mn} = E_{(m,n)}Z\tilde{g}$, where $E_{(m,n)}(t,\omega) = e^{2\pi i mt}\, e^{2\pi i n\omega}$. It therefore follows from the biorthonormality of $\{g_{mn}\}$ and $\{\tilde{g}_{mn}\}$ that $Z\tilde{g} = 1/\overline{Zg}$. Therefore, using (3.5) we formally compute

$$\begin{aligned} \overline{ZP\tilde{g}(t,\omega)} &= \frac{t}{Zg(t,\omega)} - \tfrac{1}{2\pi i}\, \partial_2(1/Zg)(t,\omega) \\ &= \frac{t\, Zg(t,\omega) + \frac{1}{2\pi i}\, \partial_2 Zg(t,\omega)}{Zg(t,\omega)^2} \\ &= \frac{ZPg(t,\omega)}{Zg(t,\omega)^2}. \end{aligned} \tag{3.6}$$

From Lemma 2.10, $|Zg|$ is an essentially constant function (i.e., is bounded below from zero and above from infinity). It therefore follows from (3.6) that $ZPg \in L^2(Q)$ if and only if $ZP\tilde{g} \in L^2(Q)$, from which (3.3) follows by the unitarity of Z. ∎

The formal arguments in the proof of Proposition 3.6 above are justified in [**BHW**] through a mixture of classical and distributional differentiation. We therefore pose the following problem which, if answered positively, would establish Proposition 3.6 without the use of differentiation.

PROBLEM 3.7. Given $g \in L^2(\mathbf{R})$ such that $\{g_{mn}\}$ is an exact Gabor frame. Assume $f \in L^2_{\text{loc}}(\mathbf{R})$ is such that $f \cdot \overline{g_{mn}} \in L^1(\mathbf{R})$ and $\langle f, g_{mn}\rangle = 0$ for all $m, n \in \mathbf{Z}$. Does it then follow that $f = 0$?

That Proposition 3.6 follows from a positive answer to Problem 3.7 is established as follows. Assume $\{g_{mn}\}$ is an exact Gabor frame and that $Pg \in L^2(\mathbf{R})$. Note that $P\tilde{g} \in L^2_{\text{loc}}(\mathbf{R})$ and $P\tilde{g} \cdot \overline{g_{mn}} = \tilde{g} \cdot \overline{Pg_{mn}} \in L^1(\mathbf{R})$ for all m, n. Moreover,

$$\begin{aligned}
\langle P\tilde{g}, g_{mn}\rangle &= \langle \tilde{g}, Pg_{mn}\rangle \\
&= \langle Z\tilde{g}, ZPg_{mn}\rangle \\
&= \langle 1/\overline{Zg}, E_{(m,n)}ZPg\rangle \\
&= \langle \overline{ZPg/(Zg)^2}, E_{(m,n)}Zg\rangle \\
&= \langle \overline{ZPg/(Zg)^2}, Zg_{mn}\rangle. \qquad (3.7)
\end{aligned}$$

Note that $ZPg \in L^2(Q)$ since $Pg \in L^2(\mathbf{R})$. As $|Zg|$ is essentially constant, it follows that $ZPg/(Zg)^2 \in L^2(Q)$. Therefore $h = Z^{-1}\big(\overline{ZPg/(Zg)^2}\big) \in L^2(\mathbf{R})$. From (3.7) and the unitarity of Z, we therefore have

$$\langle P\tilde{g}, g_{mn}\rangle = \langle h, g_{mn}\rangle$$

for all m, n. Hence, if the answer to Problem 3.7 is positive then $P\tilde{g} = h \in L^2(\mathbf{R})$.

We note that the analogue of Problem 3.7 with the exact Gabor frame $\{g_{mn}\}$ replaced by an arbitrary (exact) frame is false. For example, if $\{\psi_{jk}\}$ is the Haar system (cf., Section 2C), then $\psi \in L^1(\mathbf{R}) \cap L^2(\mathbf{R})$ and $\int \psi(x)\,dx = 0$. Therefore, taking $f \equiv 1$ we have $f \in L^2_{\text{loc}}(\mathbf{R})$ yet $\langle f, \psi_{jk}\rangle = 0$ for all j, k. Other examples relevant to Problem 3.7 are given in [**BHW**].

B. An uncertainty principle for exact wavelet frames. The previous section established that all exact Gabor frames suffer a severe time-frequency localization constraint. In this section we will consider whether exact wavelet frames suffer any similar constraints. We will use the notation $\psi_{jk}(x) = 2^{j/2}\,\psi(2^j x - k)$, cf., Section 3C.

First, note that the classical example of a wavelet orthonormal basis $\{\psi_{jk}\}$, i.e., the Haar system $\psi = \chi_{[0,1/2)} - \chi_{[1/2,1)}$, satisfies $\|t\,\psi(t)\|_2\,\|\gamma\,\hat{\psi}(\gamma)\|_2 = +\infty$. However, the "wavelet BLT" is false in general, as shown by the first modern example of a wavelet orthonormal basis. The *Meyer wavelet* is the function $\psi \in L^2(\mathbf{R})$ defined by $\hat{\psi}(\gamma) =$

$e^{i\gamma/2}\,\omega(|\gamma|)$, where

$$\omega(\gamma) = \begin{cases} 0, & \gamma \le \frac{1}{3} \text{ or } \gamma \ge \frac{4}{3}, \\ \sin \frac{\pi}{2} v(3\gamma - 1), & \frac{1}{3} \le \gamma \le \frac{2}{3}, \\ \cos \frac{\pi}{2} v(\frac{3\gamma}{2} - 1), & \frac{2}{3} \le \gamma \le \frac{4}{3}, \end{cases}$$

and $v \in C^\infty(\mathbf{R})$ is such that $v(\gamma) = 0$ for $\gamma \le 0$, $v(\gamma) = 1$ for $\gamma \ge 1$, $0 \le v(\gamma) \le 1$ for $\gamma \in [0,1]$, and $v(\gamma) + v(1-\gamma) = 1$ for $\gamma \in [0,1]$. That $\{\psi_{jk}\}$ is an orthonormal basis for $L^2(\mathbf{R})$ is shown in [**M**]. As $\hat{\psi} \in C_c^\infty(\mathbf{R})$, it follows that $\psi \in \mathcal{S}(\mathbf{R})$. Thus ψ has better than any polynomial localization in both time and frequency, i.e., $\|p(t)\,\psi(t)\|_2\,\|q(\gamma)\,\hat{\psi}(\gamma)\|_2 < +\infty$ for all polynomials p, q. However, ψ does not possess *exponential* localization in both time and frequency, and it is therefore natural to ask whether there exist wavelet orthonormal bases, or, more generally, exact wavelet frames, which do. Battle [**Ba2**] answered this in the negative for the case of wavelet orthonormal bases, and we now extend this result in a weak manner to exact wavelet frames.

First, however, note that while the dual frame of any Gabor frame is itself a Gabor frame, it is not always the case that the dual frame of any wavelet frame is itself a wavelet frame [**D1**]. However, this is true for *exact* wavelet frames. For, if $\{\psi_{jk}\}$ is an exact wavelet frame then there exists a unique function $\tilde{\psi} \in L^2(\mathbf{R})$ such that $\langle \psi_{jk}, \tilde{\psi} \rangle = \delta_j\,\delta_k$, namely $\tilde{\psi} = S^{-1}\psi$. It follows then that $\langle \psi_{jk}, \tilde{\psi}_{j'k'} \rangle = \delta_{jj'}\,\delta_{kk'}$, whence $\{\tilde{\psi}_{jk}\}$ is the dual frame to $\{\psi_{jk}\}$. We have then:

THEOREM 3.8. *Given $\psi \in L^2(\mathbf{R})$. If $\{\psi_{jk}\}$ is an exact wavelet frame then*

$$\|e^{|t|}\,\psi(t)\|_2\;\|e^{|\gamma|}\,\hat{\psi}(\gamma)\|_2\;\|e^{|t|}\,\tilde{\psi}(t)\|_2\;\|e^{|\gamma|}\,\hat{\tilde{\psi}}(\gamma)\|_2 \;=\; +\infty.$$

The proof of Theorem 3.8 depends on the following two lemmas.

LEMMA 3.9. *Given $\psi \in L^2(\mathbf{R})$ such that $\{\psi_{jk}\}$ is an exact wavelet frame. If $\hat{\psi} \in C(\mathbf{R}) \cap L^1(\mathbf{R}) \cap L^\infty(\mathbf{R})$ and $\hat{\tilde{\psi}} \in C(\mathbf{R}) \cap L^1(\mathbf{R}) \cap L^\infty(\mathbf{R})$ then $\hat{\psi}(0) = 0 = \hat{\tilde{\psi}}(0)$.*

PROOF: Note that $\tilde{\psi} \in C_0(\mathbf{R})$; therefore $\tilde{\psi}(x_0) \ne 0$ for some dyadic point $x_0 = 2^{-j_0}k_0$ where j_0, $k_0 \in \mathbf{Z}$ are not both zero. Given $j > j_0$, set $k_j = 2^{j-j_0}k_0$. Then

$$\begin{aligned} 0 &= \langle \psi_{jk_j}, \tilde{\psi} \rangle \\ &= \langle \hat{\psi}_{jk_j}, \hat{\tilde{\psi}} \rangle \end{aligned}$$

$$= 2^{-j/2} \int e^{-2\pi i 2^{-j}\gamma k_j} \hat{\psi}(2^{-j}\gamma) \overline{\hat{\tilde{\psi}}(\gamma)}\, d\gamma$$

$$= 2^{-j/2} \int e^{-2\pi i x_0 \gamma} \hat{\psi}(2^{-j}\gamma) \overline{\hat{\tilde{\psi}}(\gamma)}\, d\gamma. \tag{3.8}$$

Now,

$$e^{-2\pi i x_0 \gamma} \hat{\psi}(2^{-j}\gamma) \overline{\hat{\tilde{\psi}}(\gamma)} \rightarrow e^{-2\pi i x_0 \gamma} \hat{\psi}(0) \overline{\hat{\tilde{\psi}}(\gamma)}$$

pointwise as $j \to \infty$. Also,

$$|e^{-2\pi i x_0 \gamma} \hat{\psi}(2^{-j}\gamma) \overline{\hat{\tilde{\psi}}(\gamma)}| \leq \|\hat{\psi}\|_\infty |\hat{\tilde{\psi}}(\gamma)|$$

and $\hat{\tilde{\psi}} \in L^1(\mathbf{R})$, so by (3.8) and the Lebesgue Dominated Convergence Theorem,

$$\begin{aligned} 0 &= \lim_{j\to\infty} \int e^{-2\pi i x_0 \gamma} \hat{\psi}(2^{-j}\gamma) \overline{\hat{\tilde{\psi}}(\gamma)}\, d\gamma \\ &= \hat{\psi}(0) \overline{\int e^{2\pi i x_0 \gamma} \hat{\tilde{\psi}}(\gamma)\, d\gamma} \\ &= \hat{\psi}(0) \overline{\tilde{\psi}(x_0)}. \end{aligned}$$

As $\tilde{\psi}(x_0) \neq 0$ we conclude that $\hat{\psi}(0) = 0$. By symmetry, $\hat{\tilde{\psi}}(0) = 0$ as well. ∎

The following result extends Lemma 3.9 to general moments.

LEMMA 3.10. *Given $\psi \in L^2(\mathbf{R})$ such that $\{\psi_{jk}\}$ is an exact wavelet frame. If $\hat{\psi}$, $\hat{\tilde{\psi}} \in C^{N+1}(\mathbf{R}) \cap L^\infty(\mathbf{R})$ and*

$$(1+|\gamma|)^{N+1} \hat{\psi}(\gamma), \ (1+|\gamma|)^{N+1} \hat{\tilde{\psi}}(\gamma) \in L^1(\mathbf{R})$$

then

$$\int x^m \psi(x)\, dx = \int x^m \tilde{\psi}(x)\, dx = 0$$

for $m = 0, \ldots, N$.

PROOF: That $\int \psi(x)\, dx = \hat{\psi}(0) = 0$ follows from Lemma 3.9. Assume now that

$$i^m (D^m \hat{\psi})(0) = \int x^m \psi(x)\, dx = 0$$

for $m = 0, \ldots, k-1$ for some $k \leq N$, and assume $D^k \hat{\psi}(0) \neq 0$. Then

$$\hat{\psi}(\gamma) = \hat{\psi}(0) + \gamma D\hat{\psi}(0) + \cdots + \frac{\gamma^k (D^k \hat{\psi})(0)}{k!} + R_k(\gamma) = \frac{\gamma^k (D^k \hat{\psi})(0)}{k!} + R_k(\gamma),$$

where R_k is the Taylor remainder

$$R_k(\gamma) = \frac{\gamma^{k+1}\,(D^{k+1}\hat{\psi})(\xi)}{(k+1)!}$$

for some ξ between 0 and γ. As $D^k\hat{\psi}$ is continuous, there exists some dyadic point $x_0 = 2^{-j_0}k_0$ such that $D^k\hat{\psi}(x_0) \neq 0$. Given $j > j_0$, set $k_j = 2^{j-j_0}k_0$. Then, as in (3.8),

$$\begin{aligned} 0 &= \int e^{-2\pi i x_0 \gamma}\, \hat{\psi}(2^{-j}\gamma)\, \overline{\hat{\tilde{\psi}}(\gamma)}\, d\gamma \\ &= \int e^{-2\pi i x_0\gamma}\, \frac{\gamma^k\,(D^k\hat{\psi})(0)}{2^{jk}\,k!}\, \overline{\hat{\tilde{\psi}}(\gamma)}\, d\gamma \;+\; \int e^{-2\pi i x_0\gamma}\, R_k(2^{-j}\gamma)\, \overline{\hat{\tilde{\psi}}(\gamma)}\, d\gamma \\ &= I_1(j) \;+\; I_2(j). \end{aligned}$$

Now,

$$|R_k(2^{-j}\gamma) \le \frac{|\gamma|^{k+1}\,\|D^{k+1}\hat{\psi}\|_\infty}{2^{j(k+1)}\,(k+1)!} = \frac{C_1\,|\gamma|^{k+1}}{2^{j(k+1)}} \le \frac{C_1\,(1+|\gamma|)^{N+1}}{2^{j(k+1)}}.$$

Thus

$$I_2(j) \le \frac{C_1}{2^{j(k+1)}} \int (1+|\gamma|)^{N+1}\, |\hat{\tilde{\psi}}(\gamma)|\, d\gamma = \frac{C_2}{2^{j(k+1)}},$$

so $I_2(j) \in \mathcal{O}(2^{-j(k+1)})$. Also,

$$\begin{aligned} I_1(j) &= \frac{D^k\hat{\psi}(0)}{2^{jk}\,k!} \int e^{-2\pi i x_0\gamma}\, \gamma^k\, \overline{\hat{\tilde{\psi}}(\gamma)}\, d\gamma \\ &= \frac{C_3}{2^{jk}}\, \overline{\int \gamma^k\, \hat{\tilde{\psi}}(\gamma)\, e^{2\pi i x_0\gamma}\, d\gamma} \\ &= \frac{C_3\, i^k}{2^{jk}}\, \overline{\left((i\gamma)^k\, \hat{\tilde{\psi}}(\gamma)\right)^{\vee}(x_0)} \\ &= \frac{C_3\, i^k}{2^{jk}}\, \overline{D^k\tilde{\psi}(x_0)}. \end{aligned}$$

Thus $I_1(j) \neq 0$ for all j and $I_1(j) \in \mathcal{O}(2^{-jk}) \setminus \mathcal{O}(2^{-j(k+1)})$. As $I_1(j) = -I_2(j)$, this is a contradiction. ∎

The proof of Theorem 3.8 now follows immediately.

PROOF OF THEOREM 3.8: Assume $e^{|t|}\,\psi(t)$, $e^{|\gamma|}\,\hat{\psi}(\gamma)$, $e^{|t|}\,\tilde{\psi}(t)$, $e^{|\gamma|}\,\hat{\tilde{\psi}}(\gamma) \in L^2(\mathbf{R})$. It then follows that both ψ and $\tilde{\psi} \in \mathcal{S}(\mathbf{R})$. Therefore, from Lemma 3.10, all moments of ψ and $\tilde{\psi}$ vanish, which implies $\psi = \tilde{\psi} = 0$, a contradiction. ∎

Theorem 3.8 is closer in analogy to the weak BLT, Theorem 3.5, than to the BLT, Theorem 3.3. In the Gabor case, the weak BLT and the BLT have been shown to be equivalent; we leave as an open problem whether Theorem 3.8 can be improved to a strong form which states that if $\{\psi_{jk}\}$ is an exact wavelet frame then neither ψ nor $\tilde{\psi}$ can have exponential localization both in time and frequency.

REFERENCES

[**AT**] L. Auslander and R. Tolimieri, "Abelian Harmonic Analysis, Theta Functions and Function Algebras on a Nilmanifold," Lecture Notes in Mathematics, No. 436, Springer–Verlag, New York, 1975.

[**B**] R. Balian, *Un principe d'incertitude fort en théorie du signal ou en mécanique quantique*, C. R. Acad. Sci. Paris **292** (1981), 1357–1362.

[**Ba1**] G. Battle, *Heisenberg proof of the Balian-Low theorem*, Lett. Math. Phys. **15** (1988), 175–177.

[**Ba2**] ———, *Phase space localization theorem for ondelettes*, J. Math. Phys. **30** (1989), 2195–2196.

[**BHW**] J. Benedetto, C. Heil, and D. Walnut, *Remarks on the proof of the Balian–Low theorem*, preprint.

[**C**] A. Cohen, *Ondelettes, analysis multirésolutions et filters mirroirs en quadrature*, Annales de l'institute Henri Poincaré **7** (1990), 439–459.

[**D1**] I. Daubechies, *The wavelet transform, time-frequency localization and signal analysis*, IEEE Trans. Inform. Th. **39** (1990), 961–1005.

[**D2**] ———, *Orthonormal bases of compactly suported wavelets*, Comm. Pure Appl. Math. **41** (1988), 909–996.

[**DGM**] I. Daubechies, A. Grossmann, and Y. Meyer, *Painless nonorthogonal expansions*, J. Math. Phys. **27** (1986), 1271–1283.

[**DJ**] I. Daubechies and A.J.E.M. Janssen, *Two theorems on lattice expansions*, IEEE Trans. Inform. Th., to appear.

[**DS**] R.J. Duffin and A.C. Schaeffer, *A class of nonharmonic Fourier series*, Trans. Amer. Math. Soc. **72** (1952), 341–366.

[**G**] D. Gabor, *Theory of communications*, J. Inst. Elec. Eng. (London) **93** (1946), 429–457.

[**HW**] C. Heil and D. Walnut, *Continuous and discrete wavelet transforms*, SIAM Review **31** (1989), 628–666.

[**J**] A.J.E.M. Janssen, *Bargmann transform, Zak transform, and coherent states*, J. Math. Phys. **23** (1982), 720–731.

[**L**] F. Low, *Complete sets of wave packets*, "A Passion for Physics—Essays in Honor of Geoffrey Chew," C. DeTar, et al., ed., World Scientific, Singapore (1985), 17–22.

[**Ma**] S. Mallat, *Multiresolution approximations and wavelet orthonormal bases of* $L^2(\mathbf{R})$, Trans. Amer. Math. Soc. **315** (1989), 69–87.

[**M**] Y. Meyer, *Principe d'incertitude, bases hilbertiennes et algèbres d'opérateurs*, Séminaire Bourbaki **662** (1985-86).

The MITRE Corporation
McLean, Virginia 22102

MSC 1991: Primary 42A99
Secondary 42A65, 46B15

Operator Theory:
Advances and Applications, Vol. 58

DISTRIBUTION OF ZEROS OF MATRIX-VALUED CONTINUOUS ANALOGUES OF ORTHOGONAL POLYNOMIALS

Robert L. Ellis, Israel Gohberg, and David C. Lay

The number of zeros in the upper half-plane, counting multiplicities, of a matrix-valued continuous analogue of an orthogonal polynomial is shown to equal the number of negative eigenvalues of the associated integral operator with a matrix-valued kernel. This result generalizes a theorem of Krein and Langer on scalar-valued orthogonal functions to a noncommutative case. The proof relies on properties of orthogonal operator polynomials, Toeplitz operators, and Wiener-Hopf equations.

TABLE OF CONTENTS

INTRODUCTION

In a series of papers beginning at least as early as 1955, M. G. Krein introduced and studied a class of entire functions of the form

$$\Phi_a(\lambda) = e^{ia\lambda}\left(1 + \int_0^a g(s)e^{-i\lambda s}\,ds\right). \tag{0.1}$$

Here a is a real number and g is a solution in $L_1(0,a)$ of the equation

$$g(t) - \int_0^a k(t-s)g(s)\,ds = k(t) \qquad (0 \le t \le a), \tag{0.2}$$

where k is in $L_1(-a,a)$ and $k(-t) = \overline{k(t)}$. Krein viewed the functions Φ_a as continuous analogues of the classical Szegö orthogonal polynomials. To justify this, we recall that the classical Szegö polynomials determined by a positive function h on the unit circle $\mathbb{T}$ with an absolutely convergent Fourier series $\sum_{j=-\infty}^{\infty} h_j z^j$ may be written in the form

$$p_n(z) = b_0 z^n + b_1 z^{n-1} + \cdots + b_{n-1} z + b_n = \sum_{k=0}^{n} b_k z^{n-k}, \tag{0.3}$$

where the coefficients satisfy the equation

$$\begin{pmatrix} h_0 & \cdots & h_{-n} \\ \vdots & \ddots & \vdots \\ h_n & \cdots & h_0 \end{pmatrix} \begin{pmatrix} b_0 \\ b_1 \\ \vdots \\ b_n \end{pmatrix} = \begin{pmatrix} 1 \\ 0 \\ \vdots \\ 0 \end{pmatrix} \tag{0.4}$$

The polynomials p_n $(n \geq 0)$ are orthogonal with respect to the scalar product on $L_2(\mathbb{T})$ determined by h:

$$\langle f, g \rangle = \frac{1}{2\pi} \int_0^{2\pi} \overline{g(e^{i\theta})} h(e^{i\theta}) f(e^{i\theta})\, d\theta$$

The Toeplitz matrix H that appears in (0.4) is positive definite. In the continuous case, the role of H is played by $I - K$, where K is a compact integral operator on $L_1(0,a)$ with a selfadjoint kernel $k(t-s)$ as in (0.2). The formal analogue of (0.4) is

$$f(t) - \int_0^a k(t-s) f(s)\, ds = \delta(t), \qquad (0 \leq t \leq a). \tag{0.5}$$

To avoid using the delta function we substitute $g(t) = f(t) - \delta(t)$, so that (0.5) may be rewritten as (0.2).

To obtain a continuous analogue of the Szegö polynomials p_n $(n \geq 0)$ in (0.4), we replace the integer n by a positive real number a, and p_n by a function Φ_a with support in $[0,a]$. We write $z = e^{i\lambda}$, replace z^k $(0 \leq k \leq n)$ by $e^{i\lambda t}$ $(0 \leq t \leq a)$, and integrate from 0 to a rather than sum from 0 to n. Thus the continuous analogue of p_n is given by

$$\Phi_a(\lambda) = \int_0^a f(t) e^{i\lambda(a-t)}\, dt = e^{i\lambda a} \int_0^a f(t) e^{-i\lambda t}\, dt$$

Substituting $g(t) = f(t) - \delta(t)$ as before, we have

$$\Phi_a(\lambda) = e^{i\lambda a}\left(1 + \int_0^a g(t)e^{-i\lambda t}dt\right)$$

Setting $g(t) = 0$ for t outside $[0, a]$ and introducing the Fourier transform

$$\hat{g}(\lambda) = \int_{-\infty}^{\infty} g(t)e^{i\lambda t}\,dt = \int_0^a g(t)e^{i\lambda t}\,dt,$$

we may also write

$$\Phi_a(\lambda) = e^{ia\lambda}[1 + \hat{g}(-\lambda)]. \tag{0.6}$$

The classical theorem on the zeros of Szegö orthogonal polynomials has been generalized in several ways. In [K3], M. G. Krein considered the case of (0.4) in which the weight h is not strictly positive. He showed that if p_n is the orthogonal polynomial determined by (0.3) and (0.4), then p_n has no zeros on the unit circle, and the number of zeros of p_n in the exterior of the unit disk equals the number of negative eigenvalues of the Toeplitz matrix H in (0.4). For another approach to the proof of Krein's theorem, see [EGL]. Recently, attention has been given to the case in which the weight h is an indefinite matrix- or operator-valued function. See [G] for six papers that contain several new methods and substantial generalizations: [AG], [A], [BG], [D], [GL], and [L].

For the case of a scalar weight, M. G. Krein [K1] introduced continuous analogues of orthogonal polynomials for a definite inner product. In [KL1] Krein and H. Langer considered the indefinite scalar case and stated that the entire function Φ_a in (0.1) does not vanish on the real line and has as many zeros in the lower half-plane as the operator $I - K$ determined by the left side of (0.2) has negative eigenvalues. A full proof, along with other important results and manifold applications, appears in [KL2].

The aim of the present paper is to generalize the result of Krein and Langer to the case of matrix-valued orthogonal functions. In the following theorem and elsewhere, $L_1^{m\times m}(a, b)$ denotes the space of all $m \times m$ matrix functions whose entries are in $L_1(a, b)$.

The norm of each k in $L_1^{m\times m}(a,b)$ is given by

$$\|k\| = \int_a^b \|k(t)\|\, dt,$$

where $\|k(t)\|$ is the norm of the matrix $k(t)$ as an operator on $\mathbb{C}^n$ with its Euclidean norm.

MAIN THEOREM. *Let $k \in L_1^{m\times m}(-a,a)$ with $k(-t) = k(t)^*$ for $0 \le t \le a$. Assume that there are g and h in $L_1^{m\times m}(0,a)$ that satisfy the integral equations*

$$g(t) - \int_0^a k(t-s)g(s)\, ds = k(t) \qquad (0 \le t \le a)$$

and

$$h(t) - \int_0^a h(s)k(t-s)\, ds = k(t) \qquad (0 \le t \le a),$$

respectively. Let

$$\Psi_a(\lambda) = e^{ia\lambda}\left(I + \int_0^a e^{-i\lambda s}\, g(s)\, ds\right)$$

$$\Psi_a(\lambda) = e^{ia\lambda}\left(I + \int_0^a e^{-i\lambda s}\, h(s)\, ds\right).$$

Then Φ_a and Ψ_a are invertible on the real line, and the number of zeros of $\det(\Phi_a)$ in the lower half plane and the number of zeros of $\det(\Psi_a)$ in the lower half plane, counting multiplicities, both equal the number of negative eigenvalues of the operator $I - K$, where K is defined on $L_1^{m\times m}(0,a)$ by

$$(K\varphi)(t) = \int_0^a k(t-s)\varphi(s)\, ds \qquad (0 \le t \le a).$$

Our approach to the problem is different from that of Krein and Langer. The main idea is to change the problem into one analogous to the discrete case but involving operator-valued polynomials. To do this, we partition the interval $[0,a]$ into suitably small subintervals of length τ, and "pull back" all the integration onto the single interval $[0,\tau]$. We show that the operator K above is intertwined with a "partitioned integral operator" that is represented by a Toeplitz matrix whose entries are integral operators on $[0,\tau]$. Our treatment of this partitioned integral operator is modeled after the block-matrix case studied in [GL]. Once this is done, we translate our results back to the continuous case. This procedure is not accomplished easily and involves several steps.

The paper consists of three chapters. Chapter 1 contains general definitions and preliminary results. Theorem 1.5 connects the number of negative eigenvalues of the integral operator $I - K$ with the number of zeros, in the exterior of the unit disk, of an operator polynomial whose coefficients arise from an operator equation that is analogous to (0.4). The Toeplitz matrix that appears in this operator equation represents the partitioned integral operator for $I - K$.

Chapter 2 contains general facts about operator polynomials, which are used to prove Theorem 1.5. Chapter 3 contains the proof of the matrix generalization of the Krein-Langer theorem. An outline of its proof appears in Section 3.2.

CHAPTER 1. PRELIMINARY RESULTS

1.1. Krein Functions of the First and Second Kind

Let a be a positive real number and let K be a function in $L_1^{m\times m}(-a,a)$. Then k determines an operator K on $L_1^{m\times m}(0,a)$ defined by

$$(K\varphi)(t) = \int_0^a k(t - s)\varphi(s)\,ds$$

Suppose that there is a solution $g \in L_1^{m\times m}(0,a)$ of the equation $(I-K)g = k$, that is,

$$g(t) - \int_0^a k(t - s)g(s)\,ds = k(t) \qquad (0 \le t \le a) \tag{1.1}$$

and let $g(t) = 0$ for t outside $[0,a]$. Let

$$\ell(t) = -2k(t) + g(t) - 2\int_0^t k(t - s)g(s)\,ds \qquad (0 \le t \le a). \tag{1.2}$$

and set $\ell(t) = 0$ outside $[0,a]$. Using (1.1) to substitute for $-2k(t)$, we find that

$$\ell(t) = -g(t) + 2\int_t^a k(t - s)g(s)\,ds \tag{1.3}$$

We define the *Krein functions of the first and second kind* associated with k by

$$\Phi_a(\lambda) = e^{ia\lambda}\left(I + \int_0^a e^{-i\lambda t} g(t)\, dt\right)$$

$$\Psi_a(\lambda) = e^{ia\lambda}\left(I + \int_0^a e^{-i\lambda t} \ell(t)\, dt\right)$$

The entire functions Φ_a and Ψ_a are matrix generalizations of the orthogonal functions of the first and second kind introduced in [K1]. It turns out to be easier to deal with the function $I + \hat{g}(\lambda)$ rather than the Krein function $\Phi_a(\lambda)$ itself. So we define

$$\tilde{\Phi}_a(\lambda) = e^{ia\lambda}\Phi_a(-\lambda) = I + \hat{g}(\lambda) \tag{1.4}$$

and

$$\tilde{\Psi}_a(\lambda) = e^{ia\lambda}\Psi_a(-\lambda) = I + \hat{\ell}(\lambda), \tag{1.5}$$

where $\hat{\ell}$ is the Fourier transform of the function ℓ in (1.2). Also, for any matrix-valued function f of a complex variable, we let

$$f^*(\lambda) = [f(\bar{\lambda})]^* \tag{1.6}$$

where the asterisk on the right denotes the adjoint of a matrix.

The goal of this section is to show that on the real line the values of the Krein functions of the first and second kind are invertible matrices. This will follow from an identity satisfied by $\tilde{\Phi}_a$ and $\tilde{\Psi}_a$ defined above. The following theorem generalizes Theorem 6.1 in [KL2] for the scalar case. The proof for the noncommutative case is a slight modification of the proof of the theorem in [KL2].

THEOREM 1.1. *Let $k \in L_1^{m\times m}(-a,a)$ satisfy $k(-t) = k(t)^*$ for $0 \le t \le a$, and assume that g is a solution of the equation*

$$g(t) - \int_0^a k(t-s)g(s)\, ds = k(t) \qquad (0 \le t \le a)$$

Then for any complex number λ,

$$\tilde{\Psi}_a^*(\lambda)\tilde{\Phi}_a(\lambda) + \tilde{\Phi}_a^*(\lambda)\tilde{\Psi}_a(\lambda) = 2I \tag{1.7}$$

where $\tilde{\Phi}_a$ *and* $\tilde{\Psi}_a$ *are defined as in (1.4) and (1.5), respectively.*

PROOF. From (1.4), (1.5), and (1.6) we have

$$\begin{aligned}\tilde{\Psi}_a^*(\lambda)\tilde{\Phi}_a(\lambda) + \tilde{\Phi}_a^*(\lambda)\tilde{\Psi}_a(\lambda) = 2I &+ \int_0^a e^{i\lambda t} g(t)\,dt + \int_0^a e^{-i\lambda t}\ell(t)^*\,dt \\ &+ \int_0^a\int_0^a e^{i\lambda(s-t)}\ell(t)^* g(s)\,ds\,dt + \int_0^a e^{i\lambda t}\ell(t)\,dt \\ &+ \int_0^a e^{-i\lambda t} g(t)^*\,dt \\ &+ \int_0^a\int_0^a e^{i\lambda(s-t)} g(t)^*\ell(s)\,ds\,dt. \end{aligned}\tag{1.8}$$

By reversing the order of integration and substituting u for $t - s$, we find that

$$\begin{aligned}\int_0^a\int_0^a e^{i\lambda(s-t)}\ell(t)^* g(s)\,ds\,dt &= \int_0^a\int_{-s}^{a-s} e^{-i\lambda u}\ell(u+s)^* g(s)\,du\,ds \\ &= \int_0^a\int_{-s}^{0} e^{-i\lambda u}\ell(u+s)^* g(s)\,du\,ds \\ &+ \int_0^a\int_0^{a-s} e^{-i\lambda u}\ell(u+s)^* g(s)\,du\,ds\end{aligned}$$

Reversing the order of integration in both double integrals on the right, we can rewrite the right side as

$$\int_{-a}^0\int_{-u}^a e^{-i\lambda u}\ell(u+s)^* g(s)\,ds\,du + \int_0^a\int_0^{a-u} e^{-i\lambda u}\ell(u+s)^* g(s)\,ds\,du$$

When u is replaced by $-v$ and then s is replaced by $t+v$ in the first integral, this becomes

$$\int_0^a\int_0^{a-v} e^{i\lambda v}\ell(t)^* g(t+v)\,dt\,dv + \int_0^a\int_0^{a-u} e^{-i\lambda u}\ell(u+s)^* g(s)\,ds\,du$$

Therefore

$$\begin{aligned}\int_0^a\int_0^a e^{i\lambda(s-t)}\ell(t)^* g(s)\,ds\,dt &= \int_0^a\int_0^{a-t} e^{i\lambda t}\ell(s)^* g(s+t)\,ds\,dt \\ &+ \int_0^a\int_0^{a-t} e^{-i\lambda t}\ell(s+t)^* g(s)\,ds\,dt\end{aligned}$$

Similarly

$$\int_0^a \int_0^a e^{i\lambda(s-t)} g(t)^* \ell(s)\, ds\, dt = \int_0^a \int_0^{a-t} e^{i\lambda t} g(s)^* \ell(s+t)\, ds\, dt + \int_0^a \int_0^{a-t} e^{-i\lambda t} g(s+t)^* \ell(s)\, ds\, dt$$

Therefore (1.8) can be rewritten as

$$\begin{aligned}
&\tilde{\Psi}_a^*(\lambda)\tilde{\Phi}_a(\lambda) + \tilde{\Phi}_a^*(\lambda)\tilde{\Psi}_a(\lambda) - 2I \\
&= \int_0^a e^{-i\lambda t} \left[g(t) + \ell(t) + \int_0^{a-t} [\ell(s)^* g(s+t) + g(s)^* \ell(s+t)] ds \right] dt \\
&+ \int_0^a e^{-i\lambda t} \left[g(t)^* + \ell(t)^* + \int_0^{a-t} [g^*(s+t)\ell(s) + \ell(s+t)^* g(s)] ds \right] dt
\end{aligned}$$

Therefore (1.7) holds if and only if

$$g(t) + \ell(t) + \int_0^{a-t} [\ell(s)^* g(s+t) + g(s)^* \ell(s+t)]\, ds = 0 \qquad (0 \le t \le a)$$

By (1.3) this becomes

$$\begin{aligned}
&2\int_t^a k(t-s) g(s)\, ds - \int_0^{a-t} g(s)^* g(s+t)\, ds \\
&+ 2\int_0^{a-t} \int_s^a g(u)^* k^*(s-u) g(s+t)\, du\, ds - \int_0^{a-t} g(s)^* g(s+t)\, ds \\
&+ 2\int_0^{a-t} \int_{s+t}^a g(s)^* k(s+t-u) g(u)\, du\, ds = 0 \qquad (0 \le t \le a)
\end{aligned}$$

or

$$\begin{aligned}
&\int_t^a k(t-s) g(s)\, ds - \int_0^{a-t} g(s)^* g(s+t)\, ds \\
&+ \int_0^{a-t} \int_s^a g(u)^* k(s-u)^* g(s+t)\, du\, ds \\
&+ \int_0^{a-t} \int_{s+t}^a g(s)^* k(s+t-u) g(u)\, du\, ds = 0 \qquad (0 \le t \le a)
\end{aligned} \tag{1.9}$$

But a straightforward substitution along with (1.1) and the fact that $k(-t) = k(t)^*$ implies that

$$\begin{aligned}
&\int_t^a k(t-s)g(s)\,ds - \int_0^{a-t} g(s)^* g(s+t)\,ds \\
&= \int_t^a k(t-s)g(s)\,ds - \int_t^a g(s-t)^* g(s)\,ds \\
&= \int_t^a [k(t-s) - g^*(s-t)]g(s)\,ds \\
&= -\int_t^a \int_0^a g^*(u)k^*(s-t-u)g(s)\,du\,ds \ (0 \le t \le a) \qquad (1.10)
\end{aligned}$$

Furthermore, substituting v for $s+t$, we have for $0 \le t \le a$,

$$\begin{aligned}
&\int_0^{a-t} \int_s^a g(u)^* k(s-u)^* g(s+t)\,du\,ds \\
&+ \int_0^{a-t} \int_{s+t}^a g(s)^* k(s+t-u)g(u)\,du\,ds \\
&= \int_t^a \int_{v-t}^a g(u)^* k(v-t-u)^* g(v)\,du\,dv \\
&+ \int_t^a \int_v^a g(v-t)^* k(v-u)g(u)\,du\,dv \qquad (1.11)
\end{aligned}$$

After a reversal of integration in the second integral, the right side becomes

$$\int_t^a \int_{v-t}^a g(u)^* k(v-t-u)^* g(v)\,du\,dv + \int_t^a \int_t^u g(v-t)^* k(v-u)g(u)\,dv\,du$$

By a substitution this becomes

$$\int_t^a \int_{v-t}^a g(u)^* k(v-t-u)^* g(v)\,du\,dv + \int_t^a \int_0^{u-t} g(v)^* k(v+t-u)g(u)\,dv\,du$$

Interchanging u and v in the last integral, using the fact that $k(u+t-v) = k^*(v-t-u)$, and then combining the two integrals, we convert this to

$$\int_t^a \int_0^a g(u)^* k(v-t-u)^* g(v)\,du\,dv$$

which therefore equals the left side of (1.11). Combining this fact with equality (1.10), we conclude that (1.9) holds, which proves the theorem.

COROLLARY 1.2. *Let $k \in L_1^{m\times m}(0,a)$ satisfy $k(-t) = k(t)^*$ for $0 \le t \le a$, and assume that g is a solution of the equation*

$$g(t) - \int_0^a k(t-s)g(s)\,ds = k(t) \qquad (0 \le t \le a).$$

Let

$$\tilde{\Phi}_a(\lambda) = I + \int_0^a e^{i\lambda t} g(t)\,dt$$

and

$$\tilde{\Psi}_a(\lambda) = I + \int_0^a e^{i\lambda t}\,\ell(t)\,dt$$

where

$$\ell(t) = -2k(t) + g(t) - 2\int_0^t k(t-s)g(s)\,ds$$

Then $\det \tilde{\Phi}_a(\lambda) \ne 0$ *and* $\det \tilde{\Psi}_a(\lambda) \ne 0$ *for λ real.*

PROOF. Take λ real and suppose that $\tilde{\Phi}_a(\lambda)$ is not invertible. Then there is a nonzero vector x such that $\tilde{\Phi}_a(\lambda)x = 0$. Since λ is real, $\tilde{\Phi}_a^*(\lambda) = [\tilde{\Phi}_a(\lambda)]^*$. Therefore multiplying equality (1.7) from the right by x and from the left by x^* yields

$$0 = 2x^*x \ne 0.$$

Therefore $\tilde{\Phi}_a(\lambda)$ must be invertible for all real λ. A similar proof shows that $\tilde{\Psi}_a(\lambda)$ is invertible for λ real.

1.2. Partitioned Integral Operators

Let a be a positive number, n a positive integer, and $\tau = \frac{a}{n+1}$. For any function φ with domain containing $[0,a]$ and for $0 \le p \le n$, we define a function φ_p with domain $[0,\tau]$ by

$$\varphi_p(t) = \varphi(p\tau + t) \qquad (0 \le t \le \tau)$$

Clearly, $\varphi \in L_1^{m\times m}(0,a)$ if and only if $\varphi_p \in L_1^{m\times m}(0,\tau)$ for $0 \le p \le n$. Define $S : L_1^{m\times m}(0,a) \to [L_1^{m\times m}(0,\tau)]^{n+1}$ by

$$S(\varphi) = \operatorname{col}(\varphi_p)_{p=0}^n \tag{1.12}$$

Then S is an isometric isomorphism when $[L_1^{m\times m}(0,\tau)]^{n+1}$ is normed by

$$\|\operatorname{col}(\varphi_p)_{p=0}^n\| = \sum_{p=0}^n \|\varphi_p\|$$

Similarly, for any function φ with domain containing $[0,a]\times[0,a]$ and for $0 \le p,q \le n$, we define a function $\varphi_{p,q}$ with domain $[0,\tau]\times[0,\tau]$ by

$$\varphi_{p,q}(t,u) = \varphi(p\tau + t,\ q\tau + u) \qquad (0 \le t,u \le \tau)$$

In case $\varphi(t,u) = k(t-u)$ for some function k with domain containing $[-a,a]$,

$$\varphi_{p,q}(t,u) = k_{p-q}(t-u) \qquad (0 \le t,u \le \tau)$$

For any function k in $L_1^{m\times m}(-a,a)$ we define operators K on $L_1^{m\times m}(0,a)$ and K_p on $L_1^{m\times m}(0,\tau)$, for $-n \le p \le n$, by

$$(K\varphi)(t) = \int_0^a k(t-s)\varphi(s)\,ds \qquad (0 \le t \le a) \tag{1.13}$$

and

$$(K_p\xi)(t) = \int_0^\tau k_p(t-s)\xi(s)\,ds \qquad (0 \le t \le \tau)$$

PROPOSITION 1.3. *Let $k \in L_1^{m\times m}(-a,a)$, let n be a positive integer, and let $\tau = \frac{a}{n+1}$. Then*

$$K = S^{-1}\mathcal{K}_n S, \tag{1.14}$$

where K is the operator defined in (1.13), $\mathcal{K}_n$ is the Toeplitz matrix $(K_{p-q})_{p,q=0}^n$, and S is defined by (1.12). In particular, K and $\mathcal{K}_n$ have the same eigenvalues with the same multiplicities as operators on $L_1^{m\times m}(0,a)$ and $[L_1^{m\times m}(0,\tau)]^{n+1}$, respectively.

PROOF. For any φ in $L_1^{m\times m}(0,a)$ and for $0 \le p \le n$ and $0 \le t \le \tau$, we have

$$\begin{aligned}(K\varphi)_p(t) = (K\varphi)(p\tau + t) &= \int_0^a k(p\tau + t - s)\varphi(s)\,ds \\ &= \sum_{q=0}^{n} \int_{q\tau}^{(q+1)\tau} k(p\tau + t - s)\varphi(s)\,ds \\ &= \sum_{q=0}^{n} \int_0^{\tau} k_{p-q}(t-s)\varphi_q(s)\,ds \\ &= \sum_{q=0}^{n} (K_{p-q}\,\varphi_q)(t).\end{aligned}$$

Therefore $S(K\varphi) = \mathcal{K}_n(S(\varphi))$, which implies that $K = S^{-1}\mathcal{K}_n\mathrm{S}$.

The matrix $\mathcal{K}_n$ in Proposition 1.3 determines an operator on $[L_1^{m\times m}(0,\tau)]^{n+1}$, which we call a *partitioned integral operator.*

PROPOSITION 1.4. *Let $k \in L_1^{m\times m}(-a,a)$, let n be a positive integer, and let $\tau = \frac{a}{n+1}$. Suppose that γ is a solution of the integral equation*

$$\gamma(t,u) - \int_0^a k(t-s)\gamma(s,u)\,ds = k(t-u) \qquad (0 \le t,u \le a), \tag{1.15}$$

and for any integers p and q with $0 \le p,q \le n$, let $\Gamma_{p,q}$ be the integral operator on $L_1^{m\times m}(0,\tau)$ given by

$$(\Gamma_{p,q}\xi)(t) = \int_0^{\tau} \gamma_{p,q}(t,u)\xi(u)\,du.$$

Then

$$(I - \mathcal{K}_n)\begin{pmatrix} I + \Gamma_{0,0} \\ \Gamma_{1,0} \\ \vdots \\ \Gamma_{n,0}\end{pmatrix} = \begin{pmatrix} I \\ 0 \\ \vdots \\ 0\end{pmatrix} \tag{1.16}$$

where $\mathcal{K}_n = (K_{p-q})_{p,q=0}^{n}$.

PROOF. For any ξ in $L_1^{m\times m}(0,\tau)$ and for any integers p and q with $0 \leq p,q \leq n$, it follows from (1.15) that

$$\int_0^\tau \gamma_{p,q}(t,u)\xi(u)\,du - \int_0^\tau \sum_{r=0}^{n} \int_{r\tau}^{(r+1)\tau} k_p(t-s)\gamma_{0,q}(s,u)\xi(u)\,ds\,du$$
$$= \int_0^\tau k_{p-q}(t-u)\xi(u)\,\mathrm{du}.$$

Therefore

$$(\Gamma_{p,q}\xi)(t) - \int_0^\tau \sum_{r=0}^{n} \int_0^\tau k_{p-r}(t-s)\gamma_{r,q}(s,u)\xi(u)\,ds\,du = (K_{p-q}\xi)(t)$$

Reversing the order of integration, we find that

$$\Gamma_{p,q} - \sum_{r=0}^{n} K_{p-r}\Gamma_{r,q} = K_{p-q} \qquad (0 \leq p,q \leq n)$$

from which it follows that

$$(\Gamma_{p,q})_{p,q=0}^{n} - \mathcal{K}_n (\Gamma_{p,q})_{p,q=0}^{n} = \mathcal{K}_n.$$

Taking the first column of both sides of this equation and rearranging, we obtain (1.16).

Equation (1.16) resembles the equation that defines the classical Szegö polynomials. (See e.g. equation (1.4) in [EGL]). Accordingly, it is natural to consider the polynomial

$$P(z) = z^n I + \sum_{j=0}^{n} z^j \Gamma_{n-j,0}\ ,$$

with operator coefficients arising from (1.16). (See also [A].) Let $\tilde{P}$ be the "reversed" polynomial

$$\tilde{P}(z) = I + \sum_{j=0}^{n} z^j \Gamma_{j,0}\ .$$

Suppose $I + \Gamma_{0,0}$ is invertible. Since $\Gamma_{0,0}, \ldots, \Gamma_{n,0}$ are compact, $\tilde{P}(z)$ is invertible for all but finitely many z in the interior of the unit disk. Therefore, since $P(z) = z^n\tilde{P}(z^{-1})$ for $z \neq 0, P(z)$ is invertible for all but finitely many z in the exterior of the unit disk. Each

z such that $P(z)$ is not invertible is called a *zero* of P. The multiplicity of each zero is defined as in [R]. (See also [GS] and Section $X1.9$ of [GGK].) We will denote by $\nu_e(P)$ the total number of zeros of P, counting multiplicities, of P in the exterior of the unit disk. The following theorem, whose proof will be postponed until later, relates $\nu_e(P)$ to the number $\nu_-(I-K)$ of negative eigenvalues of $I-K$.

THEOREM 1.5. *Let k be a continuous function from $[-a,a]$ into the space of $m\times m$ matrices, with $k(-t)=k(t)^*$, let n be a positive integer, and let $\tau=\frac{a}{n+1}$. Let $\mathcal{K}_n=(K_{p-q})_{p,q=0}^n$ and $\mathcal{K}'_{n-1}=(K_{p-q})_{p,q=0}^{n-1}$, where for $-n\le p\le n, K_p$ is the operator on $L_1^{m\times m}(0,\tau)$ defined by*

$$(K_p\varphi)(t)=\int_0^\tau k(p\tau+t-s)\varphi(s)\,ds \qquad (0\le t\le\tau)$$

Assume that $I-\mathcal{K}_n$ and $I-\mathcal{K}'_{n-1}$ are invertible and have the same number of negative eigenvalues. Then (1.15) has a solution $\gamma(t,s)$. For $0\le p\le n$, define $\Gamma_{p,0}$ on $L_1^{m\times m}(0,\tau)$ by

$$(\Gamma_{p,0}\varphi)(t)=\int_0^\tau \gamma(p\tau+t,s)\varphi(s)\,ds \qquad (0\le t\le\tau).$$

and set

$$P(z)=z^nI+\sum_{j=0}^n z^j\Gamma_{n-j,0}$$

Then $P(z)$ is invertible for $|z|=1$ and, counting multiplicities, the number of zeros of P in the exterior of the unit disk equals the number of negative eigenvalues of $I-K$.

Theorem 1.5 will be proved in Section 2.2 from more abstract theorems. Then, in order to establish the main theorem, it will still remain to prove that, counting multiplicities, the number of zeros of P in the exterior of the unit disk equals the number of zeros in the upper half plane of $\det(I+\hat{g})$.

The next proposition will be needed in the proof of Theorem 3.3.

PROPOSITION 1.6. *Let k be a continuous function from $[-a,a]$ into the space of $m\times m$ matrices, let n be a positive integer, and let $\tau=\frac{a}{n+1}$. Let $\mathcal{K}_n$ and $\mathcal{K}'_{n-1}$*

be the Toeplitz matrices described in Theorem 1.5. Then for n sufficiently large, $I - \mathcal{K}_n$ and $I - \mathcal{K}'_{n-1}$ have the same number of negative eigenvalues, counting multiplicities, as operators on $[L_1^{m\times m}(0,\tau)]^{n+1}$ and $[L_1^{m\times m}(0,\tau)]^{n}$, respectively. Further, if $I - K$ is invertible, then $I - \mathcal{K}_n$ is invertible for all n, and $I - \mathcal{K}'_{n-1}$ is invertible for all n sufficiently large.

PROOF. Throughout the proof, all eigenvalues will be counted according to multiplicities. As in (1.12), we may define an isomorphism $S_2 = S_2(n)$ from $L_2^{m\times m}(0,a)$ onto the Cartesian product $[L_2^{m\times m}(0,\tau)]^{n+1}$ by

$$S_2(\varphi) = \operatorname{col}(\varphi_j)_{j=0}^{n}.$$

The norm

$$\|\operatorname{col}(\varphi_j)_{j=0}^{n}\| = \left(\textstyle\sum_{j=0}^{n} \|\varphi_j\|^2\right)^{1/2}$$

makes S_2 into an isometry. If we consider K as an operator on $L_2^{m\times m}(0,a)$ and $\mathcal{K}_n$ as an operator on $[L_2^{m\times m}(0,\tau)]^{n+1}$, then just as in Proposition 1.3,

$$K = S_2(n)^{-1}\mathcal{K}_n S_2(n) \tag{1.17}$$

Consequently, $I - K$ and $I - \mathcal{K}_n$ have the same eigenvalues as operators on their respective L_2 spaces. Given $\lambda \neq 0$ and a positive integer j, we may write $(\lambda I - K)^j = \lambda^j I - A$ where A is an integral operator with a continous kernel (since K has a continuous kernel). It follows that the eigenfunctions and generalized eigenfunctions of K for nonzero eigenvalues are continuous and hence belong to $L_p^{m\times m}(0,a)$ for $1 \le p \le \infty$. In particular, K has the same nonzero eigenvalues with the same multiplicities whether it acts as an operator on $L_1^{m\times m}(0,a)$ or on $L_2^{m\times m}(0,a)$. It follows from (1.14) and (1.17) that the nonzero eigenvalues of $\mathcal{K}_n$ are the same whether the operator acts on $[L_1^{m\times m}(0,\tau)]^{n+1}$ or on $[L_2^{m\times m}(0,\tau)]^{n+1}$. There are equations similar to (1.14) and (1.17) relating $\mathcal{K}'_{n-1}$ and K', where

$$(K'\varphi)(t) = \int_0^{a-\tau} k(t - s)\varphi(s)\,\mathrm{ds}.$$

Therefore, the nonzero eigenvalues and multiplicities of $\mathcal{K}'_{n-1}$ are the same as an operator on $\left[L_1^{m\times m}(0,\tau)\right]^n$ or on $\left[L_2^{m\times m}(0,\tau)\right]^n$. Therefore, it suffices to prove the theorem when $\mathcal{K}'_{n-1}$ and $\mathcal{K}_n$ are viewed as operators on $[L_2^{m\times m}(0,\tau)]^n$ and $[L_2^{m\times m}(0,\tau)]^{n+1}$, respectively. Observe that

$$\begin{pmatrix} I - \mathcal{K}'_{n-1} & 0 \\ 0 & I \end{pmatrix} = I - \mathcal{K}_n + \Delta_n \tag{1.18}$$

where

$$\Delta_n = \begin{pmatrix} 0 & \Xi_n \\ \Theta_n & K_0 \end{pmatrix}$$

with

$$\Xi_n = \operatorname{col}(K_{-n+j})_{j=0}^{n-1} \text{and} \Theta_n = \operatorname{row}(K_{n-j})_{j=0}^{n-1}.$$

We claim that

$$\lim_{n\to\infty} \Delta_n = 0 \tag{1.19}$$

in $[L_2^{m\times m}(0,\tau)]^{n+1}$. Since the mapping $S_2 = S_2(n)$ is an isometry for each n, it will follow from (1.19) that $I - K + S_2(n)^{-1}\Delta_n S_2(n)$ may be made as close as desired (in norm) to $I - K$ by taking n sufficiently large. Since K is a compact operator, $I - K$ has only a finite number of negative eigenvalues, with finite multiplicities. Hence when n is large, $I - K + S_2(n)^{-1}\Delta_n S_2(n)$ will have the same number of negative eigenvalues as $I - K$, with the same multiplicities. (Cf.[K], page 213.) It follows that for such n, the operator $I - \mathcal{K}_n + \Delta_n$ will have the same number of negative eigenvalues as $I - \mathcal{K}_n$. In such a case, (1.18) shows that $I - \mathcal{K}'_{n-1}$ and $I - \mathcal{K}_n$ will have the same number of negative eigenvalues.

Next we prove (1.19), or equivalently,

$$\lim_{n\to\infty} \Xi_n = 0 \qquad \text{and} \qquad \lim_{n\to\infty} [\Theta_n \; K_0] = 0, \tag{1.20}$$

in the appropriate spaces. Let $M = \max\{\|k(t)\| : -a \le t \le a\}$ and let $\varphi_0, \ldots, \varphi_n$ be any elements of $L_2^{m\times m}(0,\tau)$. Then

$$\begin{aligned} \|\Xi_n \varphi_n\|^2 &= \|\operatorname{col}(K_{-n+j}\varphi_n)_{j=0}^{n-1}\|^2 \\ &= \sum_{j=0}^{n-1} \|K_{-n+j}\varphi_n\|^2 \le \sum_{j=0}^{n-1} \|K_{-n+j}\|^2 \|\varphi_n\|^2 \end{aligned}$$

But

$$\|K_{-n+j}\|^2 \leq \int_0^\tau \int_0^\tau \|k(-n\tau + j\tau + t - s)\|^2 \, ds \, dt \leq M^2\tau^2$$

Therefore since $a = (n+1)\tau$,

$$\|\Xi_n\varphi_n\|^2 \leq \sum_{j=0}^{n-1} M^2\tau^2\|\varphi_n\|^2 = nM^2\tau^2\|\varphi_n\|^2 = \frac{nM^2a^2}{(n+1)^2}\|\varphi_n\|^2$$

and hence

$$\|\Xi_n\| \leq \frac{Ma\sqrt{n}}{n+1}$$

This proves the first equality in (1.20). Similarly,

$$\begin{aligned}
\|[\Theta_n \; K_0]\mathrm{col}(\varphi_j)_{j=0}^n\|^2 &= \|\sum_{j=0}^n K_{n-j}\varphi_j\|^2 \\
&\leq \left(\sum_{j=0}^n \|K_{n-j}\|\|\varphi_j\|\right)^2 \leq \left(\sum_{j=0}^n \|K_{n-j}\|^2\right)\left(\sum_{j=0}^n \|\varphi_j\|^2\right) \\
&\leq \left(\sum_{j=0}^n M^2\tau^2\right)\|\mathrm{col}(\varphi_j)_{j=0}^n\|^2
\end{aligned}$$

so that

$$\|[\Theta_n \; K_0]\| \leq (n+1)M^2\tau^2 = \frac{a^2M^2}{n+1}$$

This implies the second equality in (1.20). Therefore $I - \mathcal{K}_n$ and $I - \mathcal{K}'_{n-1}$ have the same number of negative eigenvalues for n sufficiently large.

Finally, suppose that $I - K$ is invertible on $L_1^{m\times m}(0,a)$. Then $I - \mathcal{K}_n$ is invertible on $[L_1^{m\times m}(0,\tau)]^{n+1}$ by Proposition 1.3. By the first part of the proof, $I - \mathcal{K}_n$ is invertible on $[L_2^{m\times m}(0,\tau)]^{n+1}$. Since $S_2(n)$ is an isometry for each n, (1.19) shows that for n sufficiently large, the operator $I - K + S_2(n)^{-1}\Delta_n S_2(n)$ is invertible and hence $I - \mathcal{K}_n + \Delta_n$ is invertible. By (1.18), $I - \mathcal{K}'_{n-1}$ is invertible on $[L_2^{m\times m}(0,\tau)]^n$. Therefore the first part of the proof implies that $I - \mathcal{K}'_{n-1}$ is invertible on $[L_1^{m\times m}(0,\tau)]^n$ for sufficiently large n.

CHAPTER 2. ORTHOGONAL OPERATOR-VALUED POLYNOMIALS

In this chapter we prove Theorem 1.5. The proof will make use of a result concerning Stein equations for operators on a Hilbert space.

2.1. Stein Equations for Operators

THEOREM 2.1. *Let A be a bounded linear operator on a Hilbert space H whose essential spectrum lies in the open unit disk, and let $T = I - K$, where K is a compact selfadjoint operator on H. Suppose that $T - A^*TA$ is positive definite. Then A has no eigenvalues on the unit circle, T is invertible and, counting multiplicities, the number of negative eigenvalues of T equals the number of eigenvalues of A in the exterior of the unit disk.*

PROOF. Let

$$B = T - A^*TA \tag{2.1}$$

so that B is positive definite. We first prove the theorem under the assumption that the spectral radius of A is less than 1. It follows from (2.1) by induction that for $n \geq 1$,

$$T = U_n + V_n \tag{2.2}$$

where

$$U_n = \sum_{k=0}^{n-1} (A^*)^k B A^k \quad \text{and} \quad V_n = (A^*)^n T A^n.$$

Observe that for any $x \neq 0$, $\langle U_n x, x\rangle$ is positive and nondecreasing as a function of n and that $< V_n x, x >$ tends to zero because the spectral radius of A is less than 1, which implies that $\|A^n\|$ tends to 0 as n tends to ∞. From this and (2.2) it follows that T is positive definite. Thus the theorem is true under the assumption that A has spectral radius less than 1.

Now consider the general case, in which the essential spectrum of A lies in the open unit disk. Then there are only finitely many eigenvalues of A in the exterior of the unit disk, each having finite multiplicity. Observe that there are no eigenvalues of A on the unit circle. Indeed, if $|\lambda| = 1$ and $Ax = \lambda x$ for some $x \neq 0$, then (2.1) implies that $< Bx, x >= 0$, which contradicts the fact that B is positive definite. Therefore, the unit circle divides the spectrum of A into two spectral sets. Using the associated Riesz projections, we decompose H into a direct sum:

$$H = H_+ \oplus H_-$$

where H_+ and H_- are invariant under A, the spectrum of $A|H_+$ lies in the open unit disk, and the spectrum of $A|H_-$ lies in the exterior of the unit disk. Furthermore, H_- is finite dimensional, and $\dim H_-$ equals the number of eigenvalues of A, counting multiplicities, in the exterior of the unit disk.

Let P_+ be the orthogonal projection of H onto H_+. Since $AH_+ \subseteq H_+$, we have

$$AP_+ = P_+AP_+ \text{ and } P_+A^* = P_+A^*P_+$$

Therefore (2.1) implies that

$$P_+TP_+ - (P_+A^*P_+)(P_+TP_+)(P_+AP_+) = P_+BP_+ \tag{2.3}$$

It follows that the restrictions of P_+TP_+, P_+AP_+, and P_+BP_+ to H_+ satisfy the hypotheses of T, A, and B in the theorem, and the spectral radius of P_+AP_+ is less than one. Therefore, by the argument above, P_+TP_+ is positive definite on H_+.

Similarly, let P_- be the orthogonal projection of H onto H_-. Then $AP_- = P_-AP_-$ and $P_-A^* = P_-A^*P_-$, so

$$P_-TP_- - (P_-A^*P_-)(P_-TP_-)(P_-AP_-) = P_-BP_- \tag{2.4}$$

The restriction of P_-AP_- to H_- is invertible, because its spectrum lies in the exterior of the unit disk. Let R be its inverse. Multiplying (2.4) by R^* from the left and R from the right we have

$$(-P_-TP_-)|_{H_-} - R^*(-P_-TP_-)R = R^*(P_-BP_-)R$$

Since $R^*(P_-BP_-)R$ is positive definite, it follows from our earlier work that the restriction of $-P_-TP_-$ to H_- is positive definite and hence that P_-TP_- is negative definite on H_-.

Now let us decompose H into an orthogonal direct sum:

$$H = L_- \oplus L_0 \oplus L_+$$

where L_-(respectively, L_+) is spanned by eigenvectors corresponding to negative (respectively, positive) eigenvalues of T, and $L_0 = \text{Ker(T)}$. This is possible since $T = I + K$, where K is compact. Observe that

$$(L_- \oplus L_0) \cap H_+ = \{0\}$$

because for x in H_+

$$\langle Tx, x\rangle \;=\; \langle TP_+x, P_+x\rangle \;=\; \langle P_+TP_+x, x\rangle$$

and we proved that P_+TP_+ is positive definite on H_+. Therefore

$$\dim(L_- \oplus L_0) \leq \text{ codim } H_+ = \dim H_- \tag{2.5}$$

Similarly,

$$(L_0 \oplus L_+) \cap H_- = \{0\}$$

because P_-TP_- is negative definite on H_-. Therefore

$$\dim L_- = \text{ codim } (L_0 \oplus L_+) \;\geq\; \dim H_- \tag{2.6}$$

From (2.5) and (2.6) we conclude that $L_0 = \{0\}$ and $\dim L_- = \dim H_-$. Since $L_0 = \text{Ker } T$ and $T = I + K$ with K compact, it follows that T is invertible. Finally, the equation $\dim L_- = \dim H_-$ means that, counting multiplicities, the number of negative eigenvalues of T equals the number of eigenvalues of A in the exterior of the unit disk.

Equation (2.1) is known as a Stein equation. The result in Theorem 2.1 is well-known for matrices. (See page 103 of [H], for example.) Theorem 2.1 can also be derived from Theorem 7.1′ in Chapter I of [DK].

2.2. Zeros of Orthogonal Operator-Valued Polynomials

Next we prove an analogue of Krein's first theorem for Toeplitz matrices whose entries are operators on a Hilbert space. The proof is modeled after the proof for the block-matrix case in Theorem 2.1 of [GL]. Given a Hilbert space H and a positive integer n, we write H^n for the Hilbert space of n-tuples of vectors in H, with the standard inner product derived from H. In the next theorem, we consider operators on H^n determined in the obvious way by matrices whose entries are operators on H.

THEOREM 2.2. *Let H be a Hilbert space and let A_k $(-n \leq k \leq n)$ be operators on H with $A_0 - I, A_1, \dots, A_n$ compact and $A_{-k} = A_k^*$ for $0 \leq k \leq n$. Let T_n and T_{n-1} be the selfadjoint Toeplitz matrices $(A_{p-q})_{p,q=0}^n$ and $(A_{p-q})_{p,q=0}^{n-1}$, respectively, and assume that T_n and T_{n-1} determine invertible operators on the Hilbert spaces H^n and H^{n-1}, respectively, such that T_n and T_{n-1} have the same number of negative eigenvalues. Let $X = col(X_j)_{j=0}^n$ be the (unique) solution of the equation*

$$T_n X = col(\delta_{0j} I)_{j=0}^n \tag{2.7}$$

and define the operator polynomial P by

$$P(z) = \sum_{j=0}^{n} z^j X_{n-j}$$

Then X_0 is positive definite, $P(z)$ is invertible for z on the unit circle and, counting multiplicities, the number of zeros of P in the exterior of the unit disk equals the number of negative eigenvalues of T_n.

PROOF. The operator X exists because T_n is invertible. Since T_{n-1} is also invertible, we may write $H^n = H \times H^{n-1}$ and consider the associated factorization

$$T_n = \begin{pmatrix} A_0 & \xi^* \\ \xi & T_{n-1} \end{pmatrix} = \begin{pmatrix} I & \xi^* T_{n-1}^{-1} \\ 0 & I \end{pmatrix} \begin{pmatrix} A_0 - \xi^* T_{n-1}^{-1}\xi & 0 \\ 0 & T_{n-1} \end{pmatrix} \begin{pmatrix} I & 0 \\ T_{n-1}^{-1}\xi & I \end{pmatrix} \tag{2.8}$$

where

$$\xi = \mathrm{col}(A_1, \dots, A_n)$$

From the hypothesis that T_n and T_{n-1} have the same number of negative eigenvalues, it follows that $A_0 - \xi^* T_{n-1}^{-1}\xi$ is positive definite. Moreover, it follows from (2.8) and (2.7) that

$$\begin{aligned}\begin{pmatrix} A_0 - \xi^* T_{n-1}^{-1}\xi & 0 \\ 0 & T_{n-1}\end{pmatrix}\begin{pmatrix} I & 0 \\ T_{n-1}^{-1}\xi & I\end{pmatrix}\operatorname{col}(X_j)_{j=0}^n &= \begin{pmatrix} I & -\xi^* T_{n-1}^{-1} \\ 0 & I\end{pmatrix}\operatorname{col}(\delta_{0j}I)_{j=0}^n \\ &= \operatorname{col}(\delta_{0j}I)_{j=0}^n\end{aligned}$$

Therefore

$$X_0 = (A_0 - \xi^* T_{n-1}^{-1}\xi)^{-1}$$

and hence X_0 is positive definite.

Equation (2.7) implies that

$$A_0 X_0 \; + \; \operatorname{row}(A_{-j})_{j=1}^n \cdot \operatorname{col}(X_j)_{j=1}^n = I \tag{2.9}$$

and

$$\operatorname{col}(A_j)_{j=1}^n = -T_{n-1}\operatorname{col}(X_j X_0^{-1})_{j=1}^n \tag{2.10}$$

Taking adjoints in (2.10) and noting that T_{n-1} and X_0^{-1} are selfadjoint, we obtain

$$\operatorname{row}(A_{-j})_{j=1}^n = -\operatorname{row}(X_0^{-1}X_j^*)_{j=1}^n T_{n-1} \tag{2.11}$$

Substituting (2.11) into (2.9) yields

$$A_0 - \; \operatorname{row}(X_0^{-1}X_j^*)_{j=1}^n T_{n-1}\operatorname{col}(X_j X_0^{-1})_{j=1}^n = X_0^{-1} \tag{2.12}$$

Let K_X be the companion matrix

$$K_X = \begin{pmatrix} -X_1X_0^{-1} & I & \cdots & 0 \\ -X_2X_0^{-1} & 0 & & 0 \\ \vdots & & & \vdots \\ & & & I \\ -X_nX_0^{-1} & 0 & \cdots & 0 \end{pmatrix}$$

and let Δ be the $(n-1) \times n$ matrix defined by

$$\Delta = \begin{pmatrix} I & & 0 & 0 \\ & \ddots & & \\ 0 & & I & 0 \end{pmatrix}$$

Then

$$K_X^* T_{n-1} K_X = \left(\begin{array}{c|c} \mathrm{row}\,(X_0^{-1} X_j^*)_{j=1}^n \, T_{n-1} \mathrm{col}\,(X_j X_0^{-1})_{j=1}^n & -\mathrm{row}\,(X_0^{-1} X_j^*)\, T_{n-1}\Delta^* \\ \hline -\Delta T_{n-1}\mathrm{col}(X_j X_0^{-1})_{j=1}^n & \Delta T_{n-1}\Delta^* i \end{array}\right) \tag{2.13}$$

Since $\Delta T_{n-1}\Delta^* = T_{n-2}$, we find from (2.10) and (2.11) - (2.13) that

$$K_X^* T_{n-1} K_X = \left(\begin{array}{c|ccc} A_0 - X_0^{-1} & A_{-1} & \cdots & A_{-n+1} \\ \hline A_1 & & & \\ \vdots & & T_{n-2} & \\ A_{n-1} & & & \end{array}\right)$$

Therefore T_{n-1} satisfies the equation

$$T_{n-1} - K_X^* T_{n-1} K_X = \mathrm{diag}(X_0^{-1}, 0, \cdots, 0) \tag{2.14}$$

Solving (2.14) for T_{n-1} and substituting repeatedly back into (2.14), it follows that

$$T_{n-1} - (K_X^*)^n T_{n-1} K_X^n = \sum_{j=0}^{n-1} (K_X^*)^j R K_X^j \tag{2.15}$$

where

$$R = \mathrm{diag}(X_0^{-1}, 0, \cdots, 0)$$

Let

$$\psi^{(j)} = \mathrm{col}(\delta_{jk} I)_{k=0}^{n-1} \quad \text{and} \quad \phi^{(j)} = \mathrm{row}(\delta_{jk} I)_{k=0}^{n-1} \qquad (j = 0, \cdots, n-1)$$

Then

$$R = \psi^{(0)} X_0^{-1} \phi^{(0)}$$

Let

$$D = \begin{pmatrix} 0 & \cdots & I \\ \vdots & & \vdots \\ I & \cdots & 0 \end{pmatrix}$$

$$C_X = \begin{pmatrix} 0 & I & \cdots & 0 \\ \vdots & & \ddots & \vdots \\ 0 & & & I \\ -X_0^{-1} X_n^* & & \cdots & -X_0^{-1} X_1^* \end{pmatrix} \tag{2.16}$$

and

$$S_x = \begin{pmatrix} X_{n-1}X_0^{-1} & \cdots & X_1X_0^{-1} & I \\ \vdots & & & \vdots \\ X_1X_0^{-1} & & & 0 \\ I & \cdots & & 0 \end{pmatrix} \tag{2.17}$$

Then by direct computation,

$$S_X^* C_X = C_X^T S_X^*$$

so that

$$C_X = (S_X^*)^{-1} C_X^T S_X^* \tag{2.18}$$

From (2.16) - (2.18) and the easily verified facts that $K_X^* = DC_XD$ and $(C_X^T)^j\psi^{(0)} = \psi^{(j)}$ for $0 \leq j \leq n-1$, it follows that

$$\begin{aligned}(K_X^*)^j\psi^{(0)} &= DC_X^jD\psi^{(0)} = D(S_X^*)^{-1}(C_X^T)^jS_X^*D\psi^{(0)} \\ &= D(S_X^*)^{-1}(C_X^T)^j\psi^{(0)} = D(S_X^*)^{-1}\psi^{(j)}\end{aligned}$$

Therefore

$$(K_X^*)^j\psi^{(0)} = D(S_X^*)^{-1}\psi^{(j)} \qquad (0 \leq j \leq n-1) \tag{2.19}$$

Taking adjoints, we find that

$$\phi^{(0)}K_X^j = \phi^{(j)}S_X^{-1}D \qquad (0 \leq j \leq n-1) \tag{2.20}$$

Using the fact that $R = \psi^{(0)}X_0^{-1}\phi^{(0)}$ and substituting (2.19) and (2.20) into (2.15), we find that

$$T_{n-1} - (K_X^*)^n T_{n-1} K_X^n = D(S_X^*)^{-1}\left(\sum_{j=0}^{n-1}\psi^{(j)}X_0^{-1}\phi^{(j)}\right)S_X^{-1}D$$

Therefore T_{n-1} satisfies the Stein equation

$$T_{n-1} - (K_X^*)^n T_{n-1} K_X^n = \Gamma^* \operatorname{diag}(X_0^{-1}, \cdots, X_0^{-1})\Gamma \tag{2.21}$$

where

$$\Gamma = \begin{pmatrix} I & \cdots & 0 & 0 \\ X_1X_0^{-1} & & & 0 \\ \vdots & & & \vdots \\ X_{n-1}X_0^{-1} & \cdots & X_1X_0^{-1} & I \end{pmatrix}^{-1}$$

From the hypotheses of the theorem it follows that $T_n - I$ is compact. Therefore $T_n^{-1} - I$ is also compact. Since X is the first column of T_n^{-1}, we deduce that $X_0 - I, X_1, \cdots$, and X_n are all compact. Therefore

$$K_X = K + S$$

where K is compact and

$$S = \begin{pmatrix} 0 & I & & \cdots & 0 \\ & 0 & I & & \vdots \\ & & & \ddots & \\ & & & & I \\ & & & & 0 \end{pmatrix}$$

Since $S^n = 0$, it follows that K_X^n is compact and hence that the essential spectrum of $K_X^n = \{0\}$. We may therefore apply Theorem 2.1 to (2.21) with $T = T_{n-1}$ and $A = K_X^n$. We conclude that K_X^n has no eigenvalues on the unit circle and the number of its eigenvalues in the exterior of the unit disk equals the number of negative eigenvalues of T_{n-1}. Therefore the same is true for K_X. Since the eigenvalues of the companion matrix K_X coincide with the zeros of the polynomial P, with the same multiplicities (cf. Example 1.1.4 in [R]), the proof of the theorem is complete.

PROOF OF THEOREM 1.5. Equation (1.15) has the form of equation (2.7), with $T_n = I - \mathcal{K}_n$ and $X = \operatorname{col}(\Gamma_{j,0} + \delta_{0j} I)_{j=0}^{n}$. Thus the polynomial P in Theorem 1.5 becomes the polynomial P in Theorem 2.2. However, in order to use Theorem 2.2, we need to convert the hypotheses and conclusions in Theorem 1.5 from statements about L_1 spaces to statements about L_2 (Hilbert) spaces.

In the proof of Proposition 1.6, we used the fact that the operator K has a continuous kernel to show that the nonzero eigenvalues of K are the same (and have the same multiplicities) whether K is considered as an operator on $L_1^{m\times m}(0,a)$ or on $L_2^{m\times m}(0,a)$. And we showed that similar statements hold for $\mathcal{K}_n$ and $\mathcal{K}'_{n-1}$. Therefore the hypotheses for Theorem 1.5 imply that the hypotheses of Theorem 2.2 are satisfied. The conclusions of Theorem 2.2 in this case are about the zeros of the polynomial $P(z)$ as an operator on the space $L_2^{m\times m}(0,\tau)$. It remains to show that these zeros are located in the same places, with the same multiplicities, when $P(z)$ is regarded as an operator on the space $L_1^{m\times m}(0,\tau)$. It turns out to be easier to work with the reversed polynomial

$$\tilde{P}(z) = I + \Gamma_{0,0} + z\Gamma_{1,0} + \cdots + z^n\Gamma_{n,0}$$

By Theorem 2.2, $I + \Gamma_{0,0}$ is positive definite on $L_2^{m\times m}(0,\tau)$ and hence $\tilde{P}(0)$ is invertible there. By Theorem 2.6 of [KMR], the zeros of $\tilde{P}$ in the open unit disk are the reciprocals of the zeros of P, with the same multiplicities, considered as operators on $L_2^{m\times m}(0,\tau)$. The same is true of $\tilde{P}$ and P on $L_1^{m\times m}(0,\tau)$(by the same theorem) because $\tilde{P}(0)$ is invertible on $L_1^{m\times m}(0,\tau)$. The invertibility of $\tilde{P}(0)$ on $L_1^{m\times m}(0,\tau)$ follows from the general fact, proved below, that the zeros of $\tilde{P}$ are the same whether the operators $\tilde{P}(z)$ act on $L_2^{m\times m}(0,\tau)$ or on $L_1^{m\times m}(0,\tau)$.

To verify the preceeding assertion, we first observe that the integral operators $\Gamma_{p,0}$ that appear in the coefficients of $\tilde{P}(z)$ have continuous kernels, since the solution $\gamma(t,s)$ of (1.14) is obviously continuous (k being continuous) and the operators $\Gamma_{p,0}$ are formed from Γ. Next, let z_0 be a zero of $\tilde{P}$ for either $L_1^{m\times m}(0,\tau)$ or $L_2^{m\times m}(0,\tau)$, and expand $\tilde{P}(z)$ in powers of $\zeta = z - z_0$:

$$\tilde{P}(z) = I - C_0 - \zeta C_1 + \cdots - \zeta^n C_n$$

where $C_0, \dots, C_n$ are all compact integral operators with continuous kernels. Let f_0 be in the kernel of $\tilde{P}(z_0)$, and let $f_1, \dots, f_r$ be an associated "Jordan chain". That is, suppose that $(I - C_0)f_0 = 0$ and

$$(I - C_0)f_k - \sum_{j=1}^{k} C_j f_{k-j} = 0 \qquad (1 \le k \le r)$$

Then

$$f_k = \sum_{j=0}^{k} C_j f_{k-j} \qquad (0 \le k \le r)$$

Since the C_j have continuous kernels, it follows that the functions $f_o, \ldots, f_r$ are continuous and hence are in both $L_2^{m\times m}(0,\tau)$ and $L_1^{m\times m}(0,\tau)$. Thus $\tilde{P}(z_0)$ has the same Jordan chains whether viewed as an operator on $L_2^{m\times m}(0,\tau)$ or on $L_1^{m\times m}(0,\tau)$. Since such chains may be used to determine the multiplicity of a zero of an operator polynomial, we conclude that $\tilde{P}$ has the same zeros, whether for each z, $\tilde{P}(z)$ is considered as an operator on $L_2^{m\times m}(0,\tau)$ or on $L_1^{m\times m}(0,\tau)$. From the earlier discussion about the zeros of P and $\tilde{P}$, the same statement holds for the zeros of P, except at the point $z = 0$. Therefore the conclusions of Theorem 2.2 apply when the operator polynomials act on $L_1^{m\times m}(0,\tau)$. Hence $P(z)$ is invertible for $|z| = 1$, and

$$\nu_e(P) = \nu_-(I - \mathcal{K}_n)$$

But Proposition 1.3 implies that $\nu_-(I - K_n) = \nu_-(I - K)$. We conclude that

$$\nu_e(P) = \nu_-(I - K).$$

2.3. On Toeplitz Matrices with Operator Entries

If E is a Banach space, then $\ell_1(E)$ will denote the Banach space of all summable sequences of vectors in E, with the norm

$$\|(\xi_0, \xi_1, \cdots)\| = \sum_{j=0}^{\infty} \|\xi_j\|.$$

The next theorem concerns an operator on $\ell_1(E)$ determined by a Toeplitz matrix whose entries are bounded linear operators on E. The theorem is a generalization of a well-known theorem involving a Toeplitz matrix with scalar entries.

THEOREM 2.3. *Let E be a Banach space with the approximation property, let Q be the polynomial*

$$Q(z) = \sum_{j=0}^{n} z^j Q_j,$$

where $Q_0 - I, Q_1, \ldots, Q_n$ are compact linear operators on E, and let T_Q be the Toeplitz operator

$$T_Q = (Q_{j-k})_{j,k=0}^{\infty},$$

where $Q_j = 0$ *for* $j < 0$ *and* $j > n$. *If* $Q(z)$ *is invertible for* $|z| = 1$, *then* T_Q *is a left-invertible Fredholm operator on* $\ell_1(E)$ *and the codimension of the range of* T_Q *equals the number of zeros of* Q *in the open unit disk, counting multiplicities.*

PROOF. Clearly $Q(z)$ is invertible for z in an annulus containing the unit circle. It follows that the function $z \mapsto Q(z)^{-1}$ is analytic on an annulus containing the unit circle. Hence the coefficients in the Laurent series for $Q(z)^{-1}$ form a convergent series in the operator norm. That is, if

$$Q(z)^{-1} \equiv S(z) = \sum_{j=-\infty}^{\infty} z^j S_j \qquad (|z| = 1),$$

then

$$\sum_{j=-\infty}^{\infty} \|S_j\| < \infty. \tag{2.22}$$

For $|z| = 1$, we have

$$I = S(z)Q(z) = \left(\sum_{j=-\infty}^{\infty} z^j S_j\right)\left(\sum_{k=-\infty}^{\infty} z^k Q_k\right) = \sum_{j=-\infty}^{\infty} z^j \left(\sum_{i=-\infty}^{\infty} S_{j-i}Q_i\right).$$

The coefficient of z^j must vanish for $j \neq 0$, and hence

$$\sum_{i=-\infty}^{\infty} S_{j-i}Q_i = \delta_{j,0} I \qquad (-\infty < j < \infty).$$

For any integer k,

$$\sum_{i=-\infty}^{\infty} S_{j-k-i}Q_i = \delta_{j-k,0} I = \delta_{j,k} I \qquad (-\infty < j, k < \infty).$$

Letting $m = k + i$, we obtain

$$\sum_{m=-\infty}^{\infty} S_{j-m}Q_{m-k} = \delta_{j,k} I \qquad (-\infty < j, k < \infty).$$

Now restrict j and k to be nonnegative. Then, since $Q_m = 0$ for $m < 0$,

$$\sum_{m=0}^{\infty} S_{j-m}Q_{m-k} = \delta_{j,k} I \qquad (0 \leq j, k < \infty) \tag{2.23}$$

The condition (2.22) shows that the Toeplitz matrix $(S_{j-m})_{j,m=0}^{\infty}$ determines a bounded linear operator on $\ell_1(E)$. And (2.23) shows that this operator is a left inverse of $T_Q = (Q_{m-k})_{m,k=0}^{\infty}$. Thus T_Q is left-invertible in the algebra of operators on $\ell_1(E)$. In particular, T_Q is a semifredholm operator.

From the perturbation theory for semifredholm operators, we know that all bounded linear operators sufficiently close to T_Q are left-invertible semifredholm operators with the same index as T_Q.(See [TL], page 262.) Since E has the approximation property, we may approximate the coefficients $Q_0, \dots, Q_n$ of Q as closely as desired by operators $R_0, \dots, R_n$, respectively, such that $R_0 - I, R_1, \dots, R_n$ all have finite rank. Let

$$R(z) = \sum_{j=0}^{n} z^j R_j,$$

and let T_R be the Toeplitz matrix whose symbol is R, namely, $T_R = (R_{m-k})_{m,k=0}^{\infty}$, where $R_j = 0$ for $j < 0$ and $j > n$. Further, let $A = (A_{j-k})_{j,k \geq 0}$, where $A_1 = I$ and $A_j = 0$ for $j \neq 1$, and observe that

$$T_Q = \sum_{j=0}^{n} Q_j A^j \text{ and } T_R = \sum_{j=0}^{n} R_j A^j$$

Clearly $\|A\| = 1$ as an operator on $\ell_1(E)$, and

$$\|T_Q - T_R\| = \|\sum_{j=0}^{n}(Q_j - R_j)A^j\| \leq \sum_{j=0}^{n} \|Q_j - R_j\|$$

Hence we can choose the R_j to make T_R a left-invertible semifredholm operator with the same index as T_Q. Moreover, since $Q(z)$ is invertible for z in the compact set $|z| = 1$, we may choose the coefficients of R to make $R(z)$ invertible for $|z| = 1$.

Since the operators $R_0 - I, R_1, \dots, R_n$ all have finite rank, there exists a topological direct sum

$$E = E^{(0)} \oplus E^{(1)}$$

such that $E^{(0)}$ is a finite-dimensional subspace invariant under $R_0, \dots, R_n$, and $E^{(1)}$ is in the intersection of the null spaces of the operators $R_0 - I, R_1, \dots R_n$. [Indeed, let X be the union of the ranges of $R_0 - I, R_1, \ldots, R_n$ and let Y be the intersection of their null spaces. Since Y is closed and has finite codimension in $E, X+Y$ is closed and there exists a closed subspace W such that $E = W \oplus (X+Y)$. Also, since $X \cap Y$ is a finite dimensional subspace of Y, there exists a closed subspace Z in Y such that $X + Y = X \oplus Z$. Then $E = W \oplus X \oplus Z$, and we may take $E^{(0)} = W \oplus X$ and $E^{(1)} = Z$.]

Since $E^{(0)}$ and $E^{(1)}$ are invariant under the operator coefficients of R, they are invariant under $R(z)$ for each z. In fact, $R(z)$ is the identity operator on $E^{(1)}$ for each z. Likewise, the subspaces $\ell_1(E^{(0)})$ and $\ell_1(E^{(1)})$ of $\ell_1(E)$ consisting of all summable sequences from $E^{(0)}$ and $E^{(1)}$, respectively, are invariant under T_R, and T_R is the identity operator on $\ell_1(E^{(1)})$. So we may view T_R as the direct sum of two operators, say,

$$T_R = T^{(0)} \oplus T^{(1)},$$

where $T^{(0)}$ and $T^{(1)}$ are the restrictions of T_R to $\ell_1(E^{(0)})$ and $\ell_1(E^{(1)})$, respectively. The operator $T^{(0)}$ is represented by a Toeplitz matrix whose entries are the restrictions to $E^{(0)}$ of the entries in T_R. The symbol of $T^{(0)}$ is the polynomial, call it $R^{(0)}$, whose coefficients are the restrictions to $E^{(0)}$ of the operators $R_0, \ldots, R_n$. Since $R(z)$ is invertible for $|z| = 1$, and since $E^{(0)}$ and $E^{(1)}$ are invariant under $R(z)$, it follows that for $|z| = 1, R^{(0)}(z)$ is invertible as an operator on $E^{(0)}$.

Since $E^{(0)}$ is finite-dimensional and the symbol of $T^{(0)}$ is invertible on the unit circle, Theorem 5.1 in Section VIII.5 of [GF] shows that $T^{(0)}$ is a Fredholm operator whose index is the negative of the winding number of det $R^{(0)}(z)$ around the unit circle. That is, ind $T^{(0)} = -m$, where

$$m = \arg \left[\det R^{(0)}(e^{i\vartheta t})\right]\Big|_{\vartheta=0}^{2\pi}$$

However, $R^{(0)}$ is analytic, so this winding number is just the number of zeros of det $R^{(0)}$ in the open unit disk, counting multiplicities. By Theorem 5.1 in [GS], the latter number equals the number of zeros of the operator polynomial $R^{(0)}$ in the open unit disk, counting multiplicities. Since $R(z)$ is the identity operator on $E^{(1)}$ for each z, it follows that the operator polynomial R itself has m zeros inside the unit disk, counting multiplicities.

Since the restriction $T^{(1)}$ is the identity operator on $\ell_1(E^{(1)})$, the index of T_R must equal the index of $T^{(0)}$. This index is $-m$, as we saw in the preceding paragraph. Recall, however, that T_R is injective since it is left-invertible, so its index is the negative of the codimension of its range. Thus T_R is a Fredholm operator, and the codimension of its range equals the number of zeros of the operator polynomial R in the open unit disk. Since T_Q has the same index as T_R, we conclude that T_Q is also a Fredholm operator. By making the polynomial R sufficiently close to Q, we can insure that they have the same number of zeros, counting multiplicities (by the operator version of the theorem of Rouché, proved in [GS].) It follows that the codimension of the range of T_Q, which equals the codimension of the range of T_R, also equals the number of zeros of Q inside $|z| < 1$. This completes the proof.

CHAPTER 3. ZEROS OF MATRIX-VALUED KREIN FUNCTIONS

In order to finish our preparation for the proof of our main theorem, we need two results about Wiener-Hopf operators.

3.1. On Wiener-Hopf Operators

The first theorem is a continuous analogue of Theorem 2.3. The Toeplitz operator is replaced by the sum of the identity operator and a Wiener-Hopf operator, the symbol of the operator is a Fourier transform rather than a polynomial, and the circle is replaced by the real line.

THEOREM 3.1. *Given g in $L_1^{m\times m}(-\infty,\infty)$, let G be the Wiener-Hopf operator defined on $L_1^{m\times m}(0,\infty)$ by*

$$(G\varphi)(t) = \int_0^\infty g(t-s)\varphi(s)ds, \qquad (0 \le t < \infty) \tag{3.1}$$

and let $\hat{g}$ *be the Fourier transform of* g. *Suppose that (a) the support of* g *is contained in* $[0,\infty)$, *and (b)* $\det(I+\hat{g}(\lambda)) \neq 0$ *for* $-\infty < \lambda < \infty$. *Then* $I+G$ *is a left-invertible Fredholm operator on* $L_1^{m\times m}(0,\infty)$, *and the codimension of the range of* $I+G$ *on* $L_1^{m\times m}(0,\infty)$ *equals the number of zeros of* $\det(I+\hat{g})$ *in the upper half plane, counting multiplicities.*

PROOF. By Remark 2 in [GK, p. 229], hypothesis (b) implies that $I+G$ is a Fredholm operator, and in this case the negative of the index of $I+G$ equals the winding number of $\det(I+\hat{g}(\lambda))$ as λ varies from $-\infty$ to ∞ along the real axis. Moreover, condition (a) insures that $\hat{g}$ is analytic in the upper half plane and $I+\hat{g}(\lambda) \to I$ as $Im\ \lambda \to \infty$. Hence the winding number of $\det(I+\hat{g}(\lambda))$ is just the number of zeros in the upper half plane, counting multiplicities. Thus the theorem will be proved when we show that $I+G$ has a left inverse on $L_1^{m\times m}(0,\infty)$.

Since condition (b) implies that $I+\hat{g}(\lambda)$ has an inverse for all real λ, it follows from Wiener's theorem for $\mathbb{R}$ that $I+\hat{g}$ is invertible in the Banach algebra of all functions of the form

$$F(\lambda) = cI + \hat{f}(\lambda),$$

where $c \in \mathbb{C}$ and $f \in L_1^{m\times m}(-\infty,\infty)$, with

$$\|F\| = |c| + \int_{-\infty}^{\infty} \|f(t)\|\, dt$$

So we may write $(I+\hat{g}(\lambda))^{-1} = I - \hat{h}(\lambda)$ for some h in $L_1^{m\times m}(-\infty,\infty)$. The identity $(I-\hat{h}(\lambda))(I+\hat{g}(\lambda)) = I$ leads to

$$\hat{h}(\lambda) + \hat{h}(\lambda)\hat{g}(\lambda) = \hat{g}(\lambda),$$

and taking inverse Fourier transforms, we have

$$h(t) + \int_{-\infty}^{\infty} h(t-s)g(s)\, ds = g(t), \qquad (-\infty < t < \infty).$$

For any $v \in \mathbb{R}$, write

$$h(t-v) + \int_{-\infty}^{\infty} h(t-s-v)g(s)\, ds \;=\; g(t-v)$$

and let $u = s + v$, so that

$$h(t-v) + \int_{-\infty}^{\infty} h(t-u)g(u-v)\,du \;=\; g(t-v).$$

Since the support of g is in $[0,\infty)$,

$$h(t-v) + \int_{v}^{\infty} h(t-u)g(u-v)\,du \;=\; g(t-v). \tag{3.2}$$

Now let H be the operator on $L_1^{m\times m}(0,\infty)$ with $h(t-s)$ as its kernel. We shall prove that $I+G$ has a left inverse, namely,

$$(I-H)(I+G) = I,$$

or equivalently,

$$H\varphi + H(G\varphi) = G\varphi, \text{ for all } \varphi \text{ in } L_1^{m\times m}(0,\infty). \tag{3.3}$$

Now

$$(H\varphi)(t) + [H(G\varphi)](t) = \int_0^{\infty} h(t-v)\varphi(v)\,dv + \int_0^{\infty} h(t-u)\left[\int_0^{\infty} g(u-v)\varphi(v)\,dv\right]$$

With the order of integration changed in the second term, the expression above equals

$$\int_0^{\infty}\left[h(t-v) + \int_v^{\infty} h(t-u)g(u-v)\,du\right]\varphi(v)\,dv = \int_0^{\infty} g(t-v)\varphi(v)\,dv = (G\varphi)(t),$$

by (3.2) above. This verifies (3.3) and completes the proof.

The next theorem establishes a useful intertwining between a Wiener-Hopf operator and a partitioned integral operator represented by an infinite Toeplitz matrix with operator entries.

THEOREM 3.2. *Let G be the Wiener-Hopf operator on $L_1^{m\times m}(0,\infty)$ with kernel g, as in (3.1). Given any $\tau > 0$, define G_p on $E = L_1^{m\times m}(0,\tau)$ by*

$$(G_p\xi)(t) = \int_0^{\tau} g_p(t-u)\xi(u)\,du, \quad p = 0,1,\ldots . \qquad (0 \le t \le \tau)$$

where $g_p(t) = g(p\tau + t)$ *and* $\xi \in E$. *Then the Toeplitz operator on* $\ell_1(E)$ *given by*

$$T_{I+G} = (\delta_{p,q}I + G_{p-q})_{p,q=0}^{\infty}$$

is similar to the operator $I + G$. *In fact,*

$$S^{-1}T_{I+G}S = I + G, \tag{3.4}$$

where S *is the isometric isomorphism from* $L_1^{m\times m}(0,\infty)$ *onto* $\ell_1(E)$ *defined by* $S\varphi = (\varphi_0, \varphi_1, \cdots)$ *for* φ *in* $L_1^{m\times m}(0,\infty)$ *and* $\varphi_p(t) = \varphi(p\tau + t)$ *for* $0 \leq t \leq \tau$ *and* $p \geq 0$.

PROOF. The operator S defined in the theorem is obviously an isomorphism onto $\ell_1(E)$, and it is isometric because

$$\|S\varphi\| = \sum_{p=0}^{\infty} \|\varphi_p\| = \sum_{p=0}^{\infty} \int_0^{\tau} |\varphi_p(t)|\, dt = \|\varphi\|.$$

The equality that defines the integral operator G in Theorem 3.1 may be decomposed into a countable system of equalities, for $p = 0, 1, 2, \dots,$

$$\begin{aligned}(G\varphi)(p\tau + t) &= \sum_{q=0}^{\infty} \int_{q\tau}^{(q+1)\tau} g(p\tau + t - s)\varphi(s)\, ds \qquad (0 \leq t < \tau)\\ &= \sum_{q=0}^{\infty} \int_0^{\tau} g((p-q)\tau + t - u)\varphi(q\tau + u)\, du, \end{aligned} \tag{3.5}$$

where $s = q\tau + u$. Using the g_p and φ_p defined above, we may write (3.5) in the form

$$(G\varphi)(p\tau + t) = \sum_{q=0}^{p} \int_0^{\tau} g_{p-q}(t-u)\varphi_q(u)\, \mathrm{du}.$$

Then (3.5) may be written as a system of equations:

$$(G\varphi)_p(t) = \sum_{q=0}^{p} (G_{p-q}\varphi_q)(t), \qquad p = 0, 1, \dots . \qquad (0 \leq t \leq \tau) \tag{3.6}$$

The relation (3.4) between $I + G$ and T_{I+G} follows easily from (3.6). We remark for later use that if the support of g in Theorem 3.2 lies in $[0, a]$ for some $a > 0$ and if $\tau = \frac{a}{n+1}$ for some $n \geq 1$, then the G_p in the theorem are 0 for $p > n + 1$, and (3.6) becomes

$$(G\varphi)_p(t) = \sum_{q=b(p,n)}^{p} \int_0^{\tau} g_{p-q}(t-u)\varphi_q(u)\, du, \qquad p = 0, 1, \dots \qquad (0 \leq t \leq \tau) \tag{3.7}$$

where $b(p,n) = \max\{0, p-n-1\}$.

3.2. Proof of the Main Theorem

It will be convenient for the proof to replace the function Φ_a in the main theorem, stated in the Introduction, by $e^{ia\lambda}\Phi_a(-\lambda)$, which is the same as $I + \hat{g}(\lambda)$. Thus we count the zeros of $\det(I+\hat{g})$ in the *upper* half plane. Similar remarks apply to Ψ_a and $I + \hat{h}$.

OUTLINE OF THE PROOF. The proof for the case when the kernel k is continuous proceeds as follows:

$$\begin{aligned} \nu_-(I-K) &= \nu_e(P) && \text{(Theorem 1.5)} \\ &= \nu_i(\tilde{P}) && \text{(Theorem 2.6 in [KMR])} \\ &= \text{codim} Im T_{I+\Gamma} && \text{(Theorem 2.3, with} Q = \tilde{P}) \\ &= \text{codim} Im T_{I+G} && \text{(for n sufficiently large)} \\ &= \text{codim} Im (I+G) && \text{(Theorem 3.2)} \\ &= \nu_u\big[\det(I+\hat{g})\big] && \text{(Theorem 3.1)} \end{aligned}$$

For the general case, we show that both $\nu_-(I-K)$ and $\nu_u(I+\hat{g})$ are stable under small perturbations of k.

THEOREM 3.3. it Let $k \in L_1^{m\times m}(-a,a)$ with $k(-t) = k(t)^*$ for $0 \le t \le$ a. Assume that there are solutions g and h of the integral equations

$$g(t) - \int_0^a k(t-s)g(s)\,ds = k(t) \qquad (0 \le t \le a) \tag{3.8}$$

and

$$h(t) - \int_0^a h(s)k(t-s)\,ds = k(t) \qquad (0 \le t \le a) \tag{3.9}$$

and let $g(t) = h(t) = 0$ for t outside $[0, a]$. Then, counting multiplicities, both the number of zeros of $\det(I + \hat{g})$ in the upper half plane and the number of zeros of $\det(I + \hat{h})$ in the upper half plane equal the number of negative eigenvalues of $I - K$.

PROOF. Throughout the proof, all eigenvalues and zeros will be counted according to multiplicities. First assume that k is continuous. Then it is easy to show that g and h are continuous also. Since $k(t-s)^* = k(s-t)$, it follows from (3.8) and (3.9) that

$$g(t)^* - \int_0^a g(s)^* k(s-t)\, ds = k(-t) \qquad (0 \le t \le a) \tag{3.10}$$

and

$$h(t)^* - \int_0^a k(s-t) h(s)^*\, ds = k(-t) \qquad (0 \le t \le a) \tag{3.11}$$

Therefore Lemma 1.2 of [GH] implies that $I - K$ is invertible as an operator on $L_1^{m\times m}(0, a)$. Consequently there is a solution of the integral equation

$$\gamma(t, u) - \int_0^a k(t-s)\gamma(s, u)\, ds = k(t-u) \qquad (0 \le t, u \le a) \tag{3.12}$$

For any $\varphi \in L_1^{m\times m}(0, a)$, if we multiply (3.12) on the right by $\varphi(u)$ and integrate from 0 to a, we find that

$$\int_0^a \gamma(t, u)\varphi(u)\, du - \int_0^a \int_0^a k(t-s)\gamma(s, u)\varphi(u)\, ds\, du$$
$$= \int_0^a k(t-u)\varphi(u)\, du \qquad (0 \le t, u \le a)$$

Reversing the order of integration in the double integral, we see that

$$\Gamma\varphi - K\Gamma\varphi = K\varphi \qquad (\varphi \in L_1^{m\times m}(0, a))$$

Therefore $(I - K)(I + \Gamma) = I$, so that

$$I + \Gamma = (I - K)^{-1}$$

Thus Lemma 1.2 of [GH] implies that

$$\gamma(t, s) = g(t-s) + g(s-t)^* + \alpha(t, s) \qquad (0 \le t, s \le \tau) \tag{3.13}$$

where

$$\alpha(t,s) = \int_0^{\min(t,s)} [g(t-r)g(s-r)^* - h(a+r-t)^* h(a+r-s)]\, dr \tag{3.14}$$

Let n be a positive integer to be chosen later and let $\tau = \frac{a}{n+1}$. Recall that for $-n \le p \le n$, the operators K_p and $\Gamma_{p,0}$ are defined by

$$(K_p\varphi)(t) = \int_0^\tau k(p\tau + t - s)\varphi(s)\, ds \text{ and } (\Gamma_{p,0}\varphi)(t) = \int_0^\tau \gamma(p\tau + t, s)\varphi(s)\, ds$$

Let $\mathcal{K}_n = (K_{p-q})_{0\le p,q\le n}$ and $\mathcal{K}'_{n-1} = (K_{p-q})_{0\le p,q\le n-1}$. By Proposition 1.6, $I - \mathcal{K}_n$ and $I - \mathcal{K}'_{n-1}$ are invertible and have the same number of negative eigenvalues for n sufficiently large. Therefore the hypotheses of Theorem 1.5 are satisfied for n sufficiently large. Let

$$P(z) = z^n I + \sum_{j=0}^{n} z^j \Gamma_{n-j,0}$$

Then Theorem 1.5 implies that for n sufficiently large, $P(z)$ is invertible for $|z| = 1$ and the number of zeros of P in the exterior of the unit disk equals the number of negative eigenvalues of $I - K$. Let

$$\tilde{P}(z) = I + \sum_{j=0}^{n} z^j \Gamma_{j,0}$$

Then

$$\tilde{P}(z) = z^n P(z^{-1}) \qquad (z \neq 0)$$

Therefore $\tilde{P}(z)$ is invertible for $|z| = 1$, and since $\tilde{P}(0) = I$, the number of zeros of $\tilde{P}$ in the open unit disk equals the number of zeros of P in the exterior of the unit disk (cf. [KMR, Theorem 2.6]). It follows that for n sufficiently large, the number of zeros of $\tilde{P}$ in the open unit disk equals the number of negative eigenvalues of $I - K$. Let $T_{I+\Gamma}$ be the Toeplitz matrix defined by

$$T_{I+\Gamma} = (\delta_{p,q} I \;+\; \Gamma_{p-q,0})_{p,q\ge 0}$$

with $\Gamma_{p,0} = 0$ for $p < 0$ and for $p > n$. Then Theorem 2.3, combined with the preceding results, implies that for n sufficiently large, $T_{I+\Gamma}$ is a left-invertible Fredholm operator and the number of negative eigenvalues of $I - K$ equals of codimension of the range of $T_{I+\Gamma}$. Let T_{I+G} be the Toeplitz matrix defined by

$$T_{I+G} = (\delta_{p,q} I \ + \ G_{p-q})_{p,q \geq 0}$$

where for $p \geq 0, G_p$ is the operator on $L_1^{m \times m}(0, \tau)$ defined by

$$(G_p \varphi)(t) = \int_0^\tau g(p\tau \ + \ t \ - \ s)\varphi(s)\, ds$$

Since g vanishes outside $[0, a], G_p = 0$ for $p < 0$ and for $p > n + 1$. We will next prove that if n is sufficiently large, then T_{I+G} is a left- invertible Fredholm operator and

$$\text{codim } Im \ T_{I+\Gamma} \ = \ \text{codim } Im \ T_{I+G}$$

For this, it suffices to prove that

$$\lim_{n \to \infty} \|T_{I+\Gamma} - T_{I+G}\| = 0$$

where the norm is the norm of $T_{I+\Gamma} - T_{I+G}$ as an operator on $\ell_1(E)$ with $E \ = L_1^{m \times m}(0, \tau)$. Then it is easy to verify, as in the proof of Theorem 2.3, that

$$\|T_{I+\Gamma} - T_{I+G}\| \leq \sum_{p=0}^{n} \|\Gamma_{p,0} - G_p\| \ + \ \|G_{n+1}\|$$

So it suffices to prove

$$\lim_{n \to \infty} \left(\sum_{p=0}^{n} \|\Gamma_{p,0} \ - \ G_p\| \ + \ \|G_{n+1}\| \right) \ = \ 0 \tag{3.15}$$

Using (3.13), we find that for any $\varphi \in L_1^{m \times m}(0, \tau)$,

$$\begin{aligned} \|(\Gamma_{0,0} - G_0)\varphi\| &= \int_0^\tau \| \int_0^\tau [\gamma(t, s) - g(t \ - \ s)]\varphi(s)\, ds\| dt \\ &= \int_0^\tau \| \int_0^\tau [g(s \ - \ t)^* + \alpha(t, s)]\varphi(s)\, ds\| dt \end{aligned}$$

where $\alpha(t, s)$ was defined in (3.14). Let

$$M = \max_{0 \leq t \leq a} \|g(t)\| \ \text{ and } \ N = \max_{0 \leq t \leq a} \|h(t)\|$$

Then since g and h vanish outside $[0, a]$,

$$\begin{aligned}\|\alpha(t,s)\| &\leq \int_0^\tau [\|g(t-r)g(s-r)^*\| + \|h(a+r-t)^*h(a+r-s)\|]\,dr \\ &\leq \tau(M^2+N^2)\end{aligned}$$

Therefore

$$\begin{aligned}\|(\Gamma_{0,0}-G_0)\varphi\| &\leq \int_0^\tau\int_0^\tau [\|g(s-r)^*\| + \|\alpha(t,s)\|]\,\|\varphi(s)\|ds\,dt \\ &\leq \tau[M+\tau(M^2+N^2)]\int_0^\tau \|\varphi(s)\|\,ds\end{aligned}$$

which implies that

$$\|\Gamma_{00}-G_0\| \leq \tau[M+\tau(M^2+N^2)]. \tag{3.16}$$

Now consider $\|\Gamma_{p,0}-G_p\|$ for $1 \leq p \leq$ n. Since $g(t) = 0$ for t outside $[0,a]$, we have $g(s-p\tau-t)^* = 0$ for $0 \leq t, s \leq \tau$ and $1 \leq p \leq n$. Therefore, for $1 \leq p \leq n$ and for any $\varphi \in L_1^{m\times m}(0,a)$, we find using (3.13) that

$$\begin{aligned}\|(\Gamma_{p,0}-G_p)\varphi\| &= \int_0^\tau \|\int_0^\tau [\gamma(p\tau+t-s)-g(p\tau+t-s)]\varphi(s)\,ds\|\,dt \\ &= \int_0^\tau \|\int_0^\tau \alpha_{p,0}(t,s)\varphi(s)\,ds\|\,dt\end{aligned}$$

where for $1 \leq p \leq n$ and $0 \leq t, s \leq \tau$,

$$\alpha_{p,0}(t,s) = \int_0^{\min(p\tau+t,s)} [g(p\tau+t-s)g(s^*-r) - h(a+r-p\tau^*-t)h(a+r-s)]\,dr$$

Since

$$\begin{aligned}\|\alpha_{p,0}(t,s)\| &\leq \int_0^\tau \|g(p\tau+t-s)g(s-r)^*\| + \|h(a+r-p\tau-t)^*h(a+r-s)\|\,dr \\ &\leq \tau(M^2+N^2)\end{aligned}$$

we have

$$\begin{aligned}\|(\Gamma_{p,0}-G_p)\varphi\| &\leq \int_0^\tau\int_0^\tau \|\alpha_p(t,s)\|\|\varphi(s)\|\,ds\,dt \\ &\leq \tau^2(M^2+N^2)\int_0^\tau \|\varphi(s)\|\,ds\end{aligned}$$

Therefore

$$\|\Gamma_{p,0} - G_p\| \leq \tau^2(M^2 + N^2) \qquad (1 \leq p \leq n)$$

Since $\tau = \frac{a}{n+1}$, this implies that

$$\sum_{p=1}^{n} \|\Gamma_{p,0} - G_p\| \leq n\tau^2(M^2 + N^2) \leq \tau \cdot a(M^2 + N^2) \tag{3.17}$$

Furthermore, for any $\varphi \in L_1^{m\times m}(0,\tau)$,

$$\begin{aligned}\|G_{n+1}\varphi\| &= \int_0^\tau \| \int_0^\tau g[(n+1)\tau + t - s)]\varphi(s)\,ds\|\,dt \\ &\leq \int_0^\tau \int_0^\tau M\|\varphi(s)\|\,ds\,dt = \tau \cdot M\|\varphi\|\end{aligned}$$

so that

$$\|G_{n+1}\| \leq \tau \cdot M \tag{3.18}$$

Since $\tau = \frac{a}{n+1}$, the equality (3.15) follows from (3.16) - (3.18). Therefore for n sufficiently large, T_{I+G} is a left-invertible Fredholm operator and codim $Im\ T_{I+\Gamma}$ = codim $Im\ T_{I+G}$. Combining this with our earlier result, we have that the number of negative eigenvalues of $I - K$ equals the codimension of the range of T_{I+G}. By Corollary 1.2, Theorem 3.1 and Theorem 3.2, the latter number equals the number of zeros of $\det(I + \hat{g})$ in the upper half plane. This completes the proof of the theorem in case k is continuous.

Now consider the general case in which $k \in L_1^{m\times m}(-a,a)$. Then k can be approximated arbitrarily closely in the norm of $L_1^{m\times m}(-a,a)$ by a continuous function k_c satisfying $k_c(-t) = k_c(t)^*$ for $0 \leq t \leq a$. For such a function k_c and for any $\varphi \in L_1^{m\times m}(0,a)$ we have

$$\begin{aligned}\|(K - K_c)\varphi\| &\leq \int_0^a \int_0^a \|k(t-s) - k_c(t-s)\|\ \|\varphi(s)\|\,ds\,dt \\ &\leq \int_0^a \int_0^a \|k(t-s) - k_c(t-s)\|\,dt\|\varphi(s)\|\,ds \\ &\leq \int_0^a \int_{-s}^{a-s} \|k(t) - k_c(t)\|\,dt\|\varphi(s)\|\,ds \\ &\leq \int_{-a}^a \|k(t) - k_c(t)\|\,dt \int_0^a \|\varphi(s)\|\,ds\end{aligned}$$

Therefore

$$\|K - K_c\| \leq \|k - k_c\|$$

Thus K can be approximated arbitrarily closely by an operator with a continuous kernel k_c satisfying $k_c(-t) = k_c(t)^*$ for $0 \leq t \leq a$. Since K is compact and $I - K$ is invertible with finitely many negative eigenvalues, if k_c is chosen close enough to k, then $I - K_c$ is invertible with the same number of negative eigenvalues as $I - K$. For any such k_c, let g_c be the solution of

$$g_c(t) - \int_0^a k_c(t - s) g_c(s)\, ds = k_c(t) \qquad (0 \leq t \leq a)$$

and let $g_c(t) = 0$ for t outside $[0, a]$. Then $g_c = (I - K_c)^{-1}(k_c)$ and $g = (I - K)^{-1}(k)$. From the equality

$$g - g_c = (I - K)^{-1}(k - k_c) + [(I - K)^{-1} - (I - K_c)^{-1}][k - (k - k_c)]$$

it follows that $\|g - g_c\|$ in $L_1^{m\times m}(0, a)$ can be made arbitrarily small by choosing k_c close enough to k. Since g and g_c vanish outside $[0, a]$,

$$(\hat{g} - \hat{g}_c)(\lambda) = \int_0^a [g(t) - g_c(t)] e^{i\lambda t}\, dt$$

which implies that

$$\|\hat{g} - \hat{g}_c\| \leq a\, \|g - g_c\|$$

Therefore, since the determinant of $I + \hat{g}$ does not vanish on the real line, it is possible to choose k_c close enough to k so that the determinant of $I + \hat{g}_c$ also does not vanish on the real line. Then it follows from the analogue of Rouché's Theorem (see Theorem 2.2 and formula (1.3) in [GS]; see also [GGK]), that the number ν_u of zeros of $\det(I + \hat{g})$ in the upper half plane equals the number of zeros of $\det(I + \hat{g}_c)$ in the upper half plane. Therefore we have

$$\nu_-(I - K) = \nu_-(I - K_c) = \nu_u[\det(I + \hat{g}_c)] = \nu_u[\det(I + \hat{g})]$$

This completes the proof of the part of the theorem involving $\det(I+\hat{g})$.

Finally, we will show that the part of the theorem involving $\det(I+\hat{h})$ follows from the part about $\det(I+\hat{g})$. Let

$$\ell(t) = k(-t) = k(t)^* \qquad (-a \le t \le a)$$

and let L be the integral operator on $L_1^{m\times m}(0,a)$ corresponding to ℓ:

$$(L\varphi)(t) = \int_0^a \ell(t-s)\varphi(s)\,ds \qquad (0 \le t \le a)$$

Taking the adjoints of equations (3.8) - (3.11), we have, respectively,

$$g(t)^* - \int_0^a g^*(s)\ell(t-s)\,ds = \ell(t) \qquad (0 \le t \le a) \tag{3.19}$$

$$h(t)^* - \int_0^a \ell(t-s)h(s)^*\,ds = l(t) \qquad (0 \le t \le a) \tag{3.20}$$

$$g(t) - \int_0^a \ell(s-t)g(s)\,ds = \ell(-t) \qquad (0 \le t \le a) \tag{3.21}$$

$$h(t) - \int_0^a h(s)\ell(s-t)\,ds = \ell(-t) \qquad (0 \le t \le a) \tag{3.22}$$

Equations (3.19) - (3.22) are precisely the equations that, according to Lemma 1.2 of [GH], must have solutions h^* and g^* in order that $I-L$ be invertible as an operator on $L_1^{m\times m}(0,a)$. Thus $I-L$ is invertible if and only if $I-K$ is invertible. It follows easily that L and K have the same nonzero eigenvalues with the same multiplicities. We now apply the part of the theorem that we have proved, but we do so with k replaced by ℓ. Then g is replaced by h^*. Thus we can conclude that the number of zeros of $\det(I+h^*)$ in the upper half plane equals the number of negative eigenvalues of $I-L$, which equals the number of negative eigenvalues of $I-K$. Thus the proof will be complete if we show that $\det(I+h^*)$ and $\det(I+\hat{h})$ have the same number of zeros in the upper half plane. But

$$\begin{aligned} h^*(\lambda) &= \int_{-\infty}^{\infty} h(t)^* e^{i\lambda t}\,dt \\ &= \left(\int_{-\infty}^{\infty} h(t)e^{-i\bar{\lambda}t}dt\right)^* \\ &= \hat{h}(-\bar{\lambda})^* \end{aligned}$$

Therefore

$$\det(I + h^*(\lambda)) = \det(I + \hat{h}(-\bar{\lambda}))$$

Since λ lies in the upper half plane if and only if $-\bar{\lambda}$ lies in the upper half plane, the desired result follows easily.

REFERENCES

[AG] Alpay, D. and I. Gohberg, On orthogonal matrix polynomials, in *Orthogonal Matrix-valued Polynomials and Applications*, I. Gohberg (ed.), Operator Theory: Advances and Applications, v. **34**, Birkhäuser Verlag, Basel, 1988.

[A] Atzmon, A., n-Orthogonal operator polynomials, in *Orthogonal Matrix-valued Polynomials and Applications*, I. Gohberg (ed.), Operator Theory: Advances and Applications, v. **34**, Birkhäuser Verlag, Basel, 1988, pp. 47-63.

[BG] Ben-Artzi, A. and I. Gohberg, Extensions of a theorem of M. G. Krein on orthogonal polynomials for the nonstationary case, in *Orthogonal Matrix-valued Polynomials and Applications*, I. Gohberg (ed.), Operator Theory: Advances and Applications, v. **34**, Birkhäuser Verlag, Basel, 1988.

[DK] Daleckii, Ju. L. and M.G. Krein, *Stability of Solutions of Differential Equations in Banach Space*, Transl.Math.Monographs 43, American Mathematical Society, Providence, R.I., 1970.

[D] Dym, H., Hermitian block Toeplitz matrices, orthogonal polynomials, reproducing kernel Pontriagin spaces, interpolation and extension, in *Orthogonal Matrix-valued Polynomials and Applications*, I. Gohberg (ed.), Operator Theory: Advances and Applications, v. **34**, Birkhäuser Verlag, Basel, 1988.

[EGL] Ellis, R.L., I. Gohberg, and D.C.Lay, On two theorems of M.G. Krein concerning polynomials orthogonal on the unit circle, *Int. Equations Op. Theory* **11** (1988), 87-104.

[G] Gohberg, I., (ed.), *Orthogonal Matrix-valued Polynomials and Applications*, Operator Theory: Advances and Applications, v. **34**, Birkhäuser Verlag, Basel, 1988.

[GF] Gohberg, I. and I.A. Feldman, *Convolution Equations and Projection Methods for their Solution*, A.M.S. Transl. Math. Monographs, v. **41**, Amer. Math. Soc., Providence, 1974.

[GG] Gohberg, I. and S. Goldberg, *Basic Operator Theory*, Birkhäuser Verlag, Basel, 1981.

[GGK] Gohberg, I., S. Goldberg and M.A. Kaashoek, *Classes of Linear Operators, I.* Operator Theory: Advances and Applications, v. **49**, Birkhäuser Verlag, Basel, 1990.

[GH] Gohberg, I. and G. Heinig, On matrix-valued integral operators on a finite interval with a kernel depending on the difference of the arguments, *Rev. Roumaine Math. Pures et Appl.* **20** (1) (1975), 55-73 [Russian].

[GK] Gohberg, I. and M.G. Krein, Systems of integral equations on a half line with kernels depending on the difference of arguments, *Uspehi Mat. Nauk (N.S)* **13** (1958), no. 2 (80), 3-72 [Russian]; *A.M.S. Transl.*, Ser. 2, **14** (1960), 212-287 [English].

[GL] Gohberg, I. and L. Lerer, Matrix generalizations of M.G. Krein theorems on orthogonal polynomials, in *Orthogonal Matrix-valued Polynomials and Applications*, I. Gohberg (ed.), Operator Theory: Advances and Applications, v. **34**, Birkhäuser Verlag, Basel, 1988, pp. 137-202.

[GS] Gohberg, I. and E. Sigal, An operator generalization of the logarithmic residue theorem and Rouché's theorem, *Mat. Sb. (N.S.)* 84 (126), 607-629 (1971) [Russian]; *Math. USSR* SB. **13** (1971), 603-625 [English].

[H] Householder, A.S., *The Theory of Matrices in Numerical Analysis*, Blaisdell Publishing Co., New York, 1964.

[K] Kato, T., *Perturbation Theory for Linear Operators*, Die Grundlehren der Mathematischen Wissenschaften, v. **132**, Springer-Verlag, New York, 1966.

[K1] Krein, M.G., Continuous analogues of propositions about orthogonal polynomials on the unit circle, *Dokl. Akad. Nauk SSSR* **105** (1955), no. 4, 637-640 [Russian].

[K2] Krein, M.G., Integral equations on a half-line with kernel depending upon the difference of the arguments, *Uspehi Mat. Nauk (N.S.)* **13** (1958), no. 5 (83), 3-120 [Russian]; *A.M.S. Transl.*, Ser. 2, **22** (1962), 163-288 [English].

[K3] Krein, M.G., On the location of the roots of polynomials which are orthogonal on the unit circle with respect to an indefinite weight, *Teor. Funkcii, Funkcional. Anal. i Prilozen* **2** (1966), 131-137 [Russian].

[KL1] Krein, M.G. and H. Langer, Continuous analogues of orthogonal polynomials on the unit circle with respect to an indefinite weight, and related extension problems, *Dokl. Akad. Nauk SSSR* **258** (1981), no. 3, 537-541 [Russian]; *Soviet Math. Dokl.* **23** (1983), no. 3, 553-557 [English].

[KL2] Krein, M.G. and H. Langer, On some continuation problems wich are closely related to the theory of operators in spaces Π_x. IV: Continuous analogues of orthogonal polynomials on the unit circle with respect to an indefinite weight and related continuation problems for some classes of functions, *J. Operator Theory* **13** (1985), 299-417.

[KMR] Kaashoek, M.A., C.V.M. van der Mee, and L. Rodman, Analytic operator functions with compact spectrum. I. Spectral nodes, linearization, and equivalence, *Int. Eq. and Operator Theory* 4 (1981), 504-547.

[L] Landau, H.J., Polynomials orthogonal in an indefinite metric, in *Orthogonal Matrix-valued Polynomials and Applications*, I. Gohberg (ed.), Operator Theory: Advances and Applications, v. **34**, Birkhäuser Verlag, Basel, 1988.

[M] Markus, A.S, *Introduction to Spectral Theory of Polynomial Operator Pencils*, A.M.S. Transl. of Math. Monographs, v. **71**, American Mathematical Society, Providence, 1988. [Russian original, 1986.]

[R] Rodman, L., *An Introduction to Operator Polynomials,* Operator Theory: Advances and Applications, v. **38**, Birkhäuser, Basel, 1989.

[TL] Taylor, A.E. and D.C.Lay, *Introduction to Functional Analysis, 2nd* ed., Wiley & Sons, New York, 1980.

Robert L. Ellis
Department of Mathematics
University of Maryland
College Park, Maryland 20742

Israel Gohberg
School of Mathematical Sciences
Raymond and Beverly Sackler
Faculty of Exact Sciences
Tel Aviv University
Tel Aviv 69978 Israel

David C. Lay
Department of Mathematics
University of Maryland
College Park, Maryland 20742

Operator Theory:
Advances and Applications, Vol. 58

THE BAND EXTENSION ON THE REAL LINE AS A LIMIT OF DISCRETE BAND EXTENSIONS, II. THE ENTROPY PRINCIPLE

I. Gohberg and **M.A. Kaashoek**

In this paper it is shown that the maximum entropy principle, which identifies the band extension on the real line, may be derived from the corresponding result for operator functions on the unit circle.

0. INTRODUCTION

Let k be an $m \times m$ matrix function with entries in $L_2([-\tau, \tau])$. An $m \times m$ matrix function f with entries in $L_1(\mathbb{R}) \cap L_2(\mathbb{R})$ is called a *positive extension* of k if

(a) $f(t) = k(t)$ for $-\tau \leq t \leq \tau$,

(b) $I - \widehat{f}(\lambda)$ is a positive definite matrix for each $\lambda \in \mathbb{R}$.

Here $\widehat{f}$ denotes the Fourier transform of f. If (b) is fulfilled, then

$$(0.1) \qquad \left(I - \widehat{f}(\lambda)\right)^{-1} = I - \widehat{\gamma}(\lambda), \qquad \lambda \in \mathbb{R}$$

where γ is again an $m \times m$ matrix function with entries in $L_1(\mathbb{R}) \cap L_2(\mathbb{R})$. A positive extension f of k is called a *band extension* if the function γ in (0.1) has the following additional property:

(c) $\gamma(t) = 0$ a.e. on $\mathbb{R} \backslash [-\tau, \tau]$.

It is known (see [DG]) that the band extension may also be characterized as the unique positive extension f of k that maximizes the *entropy integral* $\mathcal{E}(f)$, where

$$(0.2) \qquad \mathcal{E}(f) = \lim_{\varepsilon \downarrow 0} \frac{1}{2\pi} \int_{-\infty}^{\infty} \frac{\log \det(I - \widehat{f}(\lambda))}{\varepsilon^2 \lambda^2 + 1} d\lambda.$$

The main aim of the present paper is to establish the above mentioned maximum entropy characterization of the band extension by reducing it to the corresponding result for the discrete case, which concerns Fourier series on the unit circle with operator coefficients. We carry out this plan under the additional condition that there exists $\delta > 0$ such that the given matrix function k has continuous extensions defined on the closed intervals $[-\delta, 0]$ and $[0, \delta]$. Our reduction is based on partitioning of operators and does not use the usual discretization of the given k. Let us remark here that the maximum entropy principle for matrix and operator functions on the unit circle is well-understood and may be derived as a corollary of the abstract maximum entropy principle appearing in the general framework of the band method ([GKW3]). However, for the continuous case there are different entropy formulas ([AK], [Be], [Ch], [DG], see also [D], [MG]), and the maximum entropy principle does not seem to follow from the abstract analogue in the band method (see [GKW3] for an example).

In Part I of this paper it was proved that the band extension on the real line (viewed as a convolution operator) may be obtained as a limit in the operator norm of block Laurent operators of which the symbols are band extensions of appropriate discrete approximations of the given matrix function k. This result is the first main ingredient in our proof of the maximum entropy principle for the real line. The second main ingredient is Theorem I.5.1 [GK1], which expresses the entropy integral in terms of the multiplicative diagonal of the corresponding block partitioned operator. Also we employ in an essential way a notion of a generalized determinant for infinite dimensional operators that was introduced in [P] (see also Section II.1 in [GK1]).

The present Part II is divided in two chapters. The first two sections of Chapter I contain preliminary material; in the other two sections of this chapter some auxiliary results about the Perelson determinant are derived. The proof of the maximum entropy principle on the real line is given in the second chapter. Two types of approximations are considered, first from the outside of the band and next from the inside.

I. PRELIMINARIES

I.1 Some notation and terminology

In this section we introduce some notation and recall some of the terminology employed in the papers [GK1], [GK2].

By $L_2^m(\mathbb{R})$ and $L_2^m([a,b])$ we denote the Hilbert spaces of square integrable $\mathbb{C}^m$-valued functions on $\mathbb{R}$ and $[a,b]$, respectively. Let $\sigma > 0$ be a positive number. By $\mathcal{B}(\sigma)$ we denote the algebra of all bounded linear operators T on $L_2^m(\mathbb{R})$ such that the σ-block partitioning of T is a $L_2^m([0,\sigma])$-block Laurent operator with symbol in the Wiener algebra on $\mathbb{T}$ over the Hilbert-Schmidt operators on $L_2^m([0,\sigma])$. For the definition of a σ-block partitioning we refer the reader to Section I.3 of [GK2] (see also Section I.5 of [GK1]). Block Laurent operators are considered in Section I.1 of [GK2], and the Wiener algebra on $\mathbb{T}$ over the Hilbert-Schmidt operators on a Hilbert space H is introduced in Section I.2 of [GK2]. The latter algebra is denoted by $\mathcal{W}(S_2, \mathbb{T}; H)$, where S_2 stands for the operator ideal of Hilbert-Schmidt operators. See also Section I.5 of [GK2] for further properties of the algebra $\mathcal{B}(\sigma)$.

An identity operator is denoted by I; from the context it should be clear on which space its acts. Let H be a Hilbert space. We write $\ell_2(H)$ for the Hilbert space of all square summable sequences $\vec{x} = (x_i)_{i=0}^{\infty}$ with elements in H. Inner product and norm on $\ell_2(H)$ are given by

$$\langle \vec{x}, \vec{y} \rangle = \sum_{j=0}^{\infty} \langle x_j, y_j \rangle,$$

$$\|\vec{x}\| = \left(\sum_{j=0}^{\infty} \|x_j\|^2 \right)^{1/2}.$$

The Hilbert space of all double infinite square summable sequences $\boldsymbol{x} = (x_i)_{i=-\infty}^{\infty}$ with elements in H is denoted by $\oplus_{-\infty}^{\infty} H$. Instead of $\ell_2(H)$ we also write $\oplus_1^{\infty} H$.

I.2 The nonstationary operator Wiener algebra and multiplicative diagonals

Let H be a Hilbert space. By $\mathcal{NSW}(H)$ we denote the set of all double infinite operator matrices $V = (V_{ij})_{i,j=-\infty}^{\infty}$ such that each entry V_{ij} is a bounded linear operator on H and

$$\text{(2.1)} \qquad \sum_{\nu=-\infty}^{\infty} \sup_{j-i=\nu} \|V_{ij}\| < \infty.$$

The set $\mathcal{NSW}(H)$ is an algebra under the usual operations of addition and multiplication. The element $E = (\delta_{ij} I)_{i,j=-\infty}^{\infty}$ is the unit in $\mathcal{NSW}(H)$. Here δ_{ij} stands for the Kronecker delta. We shall refer to $\mathcal{NSW}(H)$ as the *nonstationary operator Wiener algebra* corresponding to H.

Each $V \in \mathcal{NSW}(H)$ defines in a canonical way a bounded linear operator on $\oplus_{-\infty}^{\infty} H$ via the rule

$$V\boldsymbol{x} = \boldsymbol{y} \Longleftrightarrow \sum_{j=-\infty}^{\infty} V_{ij} x_j = y_i, \qquad i = 0, \pm 1, \pm 2, \dots .$$

Note that $\mathcal{NSW}(H)$ is closed under taking adjoints, and thus $\mathcal{NSW}(H)$ is a $*$-subalgebra of the algebra of all bounded linear operators on $\oplus_{-\infty}^{\infty} H$.

Let $V \in \mathcal{NSW}(H)$, and assume that V is positive definite as an operator on $\oplus_{-\infty}^{\infty} H$. The arguments used in Lemma II.3.2 of [GKW2] show that V factors as

$$\text{(2.2)} \qquad V = (E + U)\Delta(V)(E + U)^*$$

where $U = (U_{ij})_{i,j=-\infty}^{\infty} \in \mathcal{NSW}(H)$ and has the following properties

(α) $U_{ij} = 0$ for $i \geq j$,

(β) $E + U$ is invertible in $\mathcal{NSW}(H)$ and the (i,j)-th entry of $(E+U)^{-1} - E$ is zero for $i \geq j$,

(γ) $\Delta(V) = \big(\delta_{ij}\Delta_j(V)\big)$, where $\Delta_j(V)$ is a bounded linear operator on H for each j and δ_{ij} is the Kronecker delta.

The factorization (2.2) with the properties (α), (β) and (γ) is unique. The diagonal operator matrix $\Delta(V)$ is called the *right multiplicative diagonal* of V.

PROPOSITION 2.1. *Let $V = (V_{ij})_{i,j=-\infty}^{\infty}$ be a positive definite operator in $\mathcal{NSW}(H)$. Then the k-th diagonal entry $\Delta_k(V)$ in the right multiplicative diagonal of V is given by*

$$\text{(2.3)} \qquad \Delta_k(V) = V_{kk} - B_k V_{[k]}^{-1} C_k,$$

where the operators $V_{[k]}$, B_k *and* C_k *are defined by*

(2.4) $$V_{[k]}: \ell_2(H) \to \ell_2(H), \quad (V_{[k]}\vec{x})_i = \sum_{j=0}^{\infty} V_{k+i+1,k+j+1}x_j, \qquad i = 0,1,2,\ldots;$$

(2.5) $$B_k: \ell_2(H) \to H, \qquad B_k\vec{x} = \sum_{j=0}^{\infty} V_{k,k+j+1}x_j;$$

(2.6) $$C_k: H \to \ell_2(H), \qquad C_k x = (V_{k+i+1,k}x)_{i=0}^{\infty}.$$

PROOF. Fix $k \in \mathbb{Z}$. From the factorization (2.2) it follows that

(2.7) $$\Delta_k(V)^{-1} = (I \quad 0 \quad \cdots)\begin{pmatrix} V_{kk} & V_{k,k+1} & \cdots \\ V_{k+1,k} & V_{k+1,k+1} & \cdots \\ \vdots & \vdots & \end{pmatrix}^{-1}\begin{pmatrix} I \\ 0 \\ \vdots \end{pmatrix}.$$

In other words, $\Delta_k(V)^{-1}$ is the $(1,1)$-entry of the inverse of the operator $V_{[k-1]}$, where $V_{[k-1]}$ is defined by (2.4) (with $k-1$ in place of k). Note that the operator $V_{[k-1]}$ may be partitioned as

$$\begin{pmatrix} V_{kk} & B_k \\ C_k & V_{[k]} \end{pmatrix},$$

where $V_{[k]}$, B_k and C_k are the operators defined by (2.4)–(2.6). But then one may use a Schur complement type of argument (cf., [BGK], Remark 1.2; see also the proof of Lemma II.2.1 in [GK1]) to obtain (2.3). □

It will be convenient to consider two special cases. In what follows $V = (V_{ij})_{i,j=-\infty}^{\infty} \in \mathcal{NSW}(H)$, and we assume that V is positive definite. Recall (cf., [GK2], Section I.1) that V is a H-block Laurent operator if V_{ij} depends on the difference $i-j$ only. In other words, $V_{ij} = V_{i-j}$ for all i and j. Assume that V is such an operator. Then V has a well-defined symbol, namely

(2.8) $$V(z) = \sum_{\nu=-\infty}^{\infty} z^{\nu}V_{\nu}, \qquad |z| = 1.$$

Since V is positive definite, the operator $V(z)$ is a positive definite operator on H for each $z \in \mathbb{T}$, and hence (see [GKW2], Lemma II.1.1) the operator function $V(\cdot)$ factors as

(2.9) $$V(z) = (I + U(z))\Delta_0(I + U(z))^*, \qquad z \in \mathbb{T},$$

where Δ_0 is a positive definite operator on H, for each $z \in \mathbb{T}$ the operator $I + U(z)$ is invertible, and the operator function $U(\cdot)$ and $\left(I + U(\cdot)\right)^{-1} - I$ are analytic on $|z| > 1$ (including ∞), have the value zero at ∞ and extend to continuous functions on $|z| \geq 1$.

The factors in the factorization (2.9) are uniquely determined by the symbol $V(\cdot)$, and one refers to Δ_0 as the *right multiplicative diagonal* of $V(\cdot)$. By passing from (2.9) to the corresponding H-block Laurent operators one sees that the i-th entry in the right multiplicative diagonal of V does not depend on i and is precisely equal to the right multiplicative diagonal of the symbol of V.

Next, we assume that V is p-periodic. The latter means that $V_{i,j} = V_{i+p,j+p}$ for all i and j. If $p = 1$, then we are back to the case considered in the preceding paragraph. Let $H^{(p)}$ denote the Hilbert space direct sum of p copies of H. The fact that V is p-periodic implies that we may view V as a $H^{(p)}$-block Laurent operator. Indeed, put

$$\widetilde{V}_i = \begin{pmatrix} V_{ip,0} & \cdots & V_{ip,p-1} \\ \vdots & & \vdots \\ V_{(i+1)p-1,0} & \cdots & V_{(i+1)p-1,p-1} \end{pmatrix}, \qquad i \in \mathbb{Z},$$

and let $\widetilde{V} = (\widetilde{V}_{i-j})_{i,j=-\infty}^{\infty}$. By identifying the spaces $\oplus_{-\infty}^{\infty} H$ and $\oplus_{-\infty}^{\infty} H^{(p)}$ we see that V may be identified with $\widetilde{V}$. It follows that the right multiplicative diagonal of V may be obtained in the following way. First determine the right multiplicative diagonal Δ of the symbol of $\widetilde{V}$. Since Δ is a positive definite operator on $H^{(p)}$, the operator Δ factors as

$$\Delta = (I + A) \begin{pmatrix} \Delta_0 & & & \\ & \Delta_1 & & \\ & & \ddots & \\ & & & \Delta_{p-1} \end{pmatrix} (I + A)^*, \tag{2.10}$$

where A is a strictly upper triangular $p \times p$ operator matrix whose entries are bounded linear operators on H and the unspecified entries in the middle term of the right hand side of (2.10) are equal to the zero operator on H. By combining the factorization (2.10) with the result of the preceding paragraph we see that in the p-periodic case the right multiplicative diagonal of V is the double infinite p-periodic diagonal operator matrix whose diagonal entries $\Delta_i(V)$ are given by

$$\Delta_{kp+i}(V) = \Delta_i, \qquad i = 0, \ldots, p-1. \tag{2.11}$$

I.3 A separation lemma for the Perelson determinant

Let $A\colon L_2^m([a,b]) \to L_2^m([a,b])$ be a Hilbert-Schmidt operator. For each $h > 0$ let M_h be the averaging operator on $L_2^m([a,b])$ defined by

$$(M_h f)(t) = \int_a^b \Delta_h(t-s) f(s)\, ds, \qquad a \le t \le b,$$

where

$$\Delta_h(t) = \begin{cases} \dfrac{1}{2h} I_m, & |t| \le h, \\ 0 \quad , & |t| > h. \end{cases}$$

The operator M_h is a Hilbert-Schmidt operator, and hence M_hAM_h is a trace class operator. We say that A belongs to the *Perelson class* (notation: $A \in \mathcal{P}$) if the limit

$$\lim_{h\downarrow 0} \operatorname{tr}(M_hAM_h) \tag{3.1}$$

exists, and in that case the quantity in (3.1) is called the *Perelson trace* of A (notation: $\operatorname{TR}(A)$). If A belongs to the Perelson class, we put

$$\operatorname{DET}(I-A) := \widetilde{\det}(I-A)\exp\big(-\operatorname{TR}(A)\big). \tag{3.2}$$

Here $\widetilde{\det}(I-A)$ is the second regularized determinant of $I-A$ (see [GKr1], Section IV.2). The quantity $\operatorname{DET}(I-A)$ is called the *Perelson determinant* of $I-A$. It can be shown (see the proof of Lemma 3.1 below) that for $A \in \mathcal{P}$ we have

$$\operatorname{DET}(I-A) = \lim_{h\downarrow 0} \det(I - M_hAM_h).$$

In fact, the latter identity is the definition of the Perelson determinant as it appears in [P]. A review of the main properties of the Perelson trace and determinant is given in Section II.1 of [GK1]. We shall need the following lemma.

LEMMA 3.1. *Let K_1 and K_2 be Hilbert-Schmidt operators on $L_2^m([a,b])$ from the Perelson class such that*

$$I - K_2 \geq I - K_1 > 0. \tag{3.3}$$

Assume that $K_1 - K_2$ is a trace class operator. Then

$$\operatorname{DET}(I-K_2) \geq \operatorname{DET}(I-K_1), \tag{3.4}$$

and equality holds in (3.4) if and only if $K_1 = K_2$.

PROOF. Condition (3.3) means that $I-K_1$ and $I-K_2$ are positive definite operators and the difference K_1-K_2 is positive semidefinite. Since the averaging operator M_h is selfadjoint, $K_1 \geq K_2$ implies that $M_hK_1M_h \geq M_hK_2M_h$. Next, note that $\|M_h\| \leq 1$. To prove the latter inequality one can use the same arguments as are used in Section 10 of [LS] for scalar functions. There exists $0 < \delta < 1$ such that

$$\langle K_1 f, f\rangle \leq (1-\delta)\langle f, f\rangle, \qquad f \in L_2^m\big([a,b]\big),$$

because $I - K_1$ is positive definite. It follows that for each $f \in L_2^m([a,b])$ we have

$$\begin{aligned}\langle M_hK_1M_hf, f\rangle &\leq (1-\delta)\langle M_hf, M_hf\rangle \\ &= (1-\delta)\|M_hf\|^2 \\ &\leq (1-\delta)\langle f, f\rangle,\end{aligned}$$

which implies that $I - M_hK_1M_h$ is positive definite. So we have shown that

$$I - M_hK_2M_h \geq I - M_hK_1M_h > 0.$$

Since $M_hK_1M_h$ and $M_hK_2M_h$ are trace class operators, we may conclude that

$$\det(I - M_hK_2M_h) \geq \det(I - M_hK_1M_h). \tag{3.5}$$

Here $\det(I - A)$ is the usual trace class determinant.

It is known that $M_hf \to f$ if $h \downarrow 0$ for each $f \in L_2^m([a,b])$ (cf., [GGK], page 112). Thus (by [GKr1], Theorem III.6.3) we have for $\nu = 1,2$ that $M_hK_\nu M_h \to K_\nu$ if $h \downarrow 0$ with respect to the Hilbert-Schmidt norm. But then we can apply [GKr1], Theorem IV.2.1 to show that

$$\widetilde{\det}(I - K_\nu) = \lim_{h\downarrow 0} \widetilde{\det}(I - M_hK_\nu M_h), \qquad \nu = 1,2,$$

and hence

$$\begin{aligned}\mathrm{DET}(I - K_\nu) &= \widetilde{\det}(I - K_\nu)\exp\big(-\mathrm{TR}(K_\nu)\big)\\ &= \lim_{h\downarrow 0} \widetilde{\det}(I - M_hK_\nu M_h)\exp\big(-\mathrm{TR}(M_hK_\nu M_h)\big)\\ &= \lim_{h\downarrow 0} \det(I - M_hK_\nu M_h), \qquad \nu = 1,2.\end{aligned}$$

By combining this limit formulas with (3.5) we obtain the inequality (3.4).

Now, assume that $\mathrm{DET}(I - K_2) = \mathrm{DET}(I - K_1)$. We have to show that $K_1 = K_2$. Put $C = K_1 - K_2$. By our hypotheses C is a trace class operator and $C \geq 0$. From $I - K_1 + C = I - K_2$ we see that

$$I - K_2 = (I - K_1)\big(I + (I - K_1)^{-1}C\big),$$

where $(I - K_1)^{-1}C$ is a trace class operator. Hence, by [GK1], Proposition II.1.2,

$$\begin{aligned}\mathrm{DET}(I - K_1) &= \mathrm{DET}(I - K_2)\\ &= \mathrm{DET}(I - K_1)\,\mathrm{DET}\big(I + (I - K_1)^{-1}C\big).\end{aligned}$$

It follows (see [GK1], Proposition II.1.1) that $\det\big(I + (I - K_1)^{-1}C\big) = 1$. Since $I - K_1$ is positive definite, we conclude that

$$\det\big(I + (I - K_1)^{-1/2}C(I - K_1)^{-1/2}\big) = 1. \tag{3.6}$$

But $(I - K_1)^{-1/2}C(I - K_1)^{-1/2} \geq 0$, and thus (3.6) implies that the eigenvalues of $(I - K_1)^{-1/2}C(I - K_1)^{-1/2}$ are zero, and hence this operator must be zero. The latter implies that $C = 0$, and thus $K_1 = K_2$. □

I.4 The Perelson determinant and factorization

For $i,j = 1,\dots,N$ let K_{ij} be a Hilbert-Schmidt operator on $L_2^m([a,b])$. Consider

$$K = \begin{pmatrix} K_{11} & \cdots & K_{1N} \\ \vdots & & \vdots \\ K_{N1} & \cdots & K_{NN} \end{pmatrix}. \tag{4.1}$$

Note that K is a Hilbert-Schmidt operator on $L_2^{mN}([a,b])$. In what follows we assume that $I-K$ is positive definite. This implies that $I-K$ factors as

$$(4.2)\qquad I-K=(I-F)\begin{pmatrix} I-D_1 & & \\ & \ddots & \\ & & I-D_N \end{pmatrix}(I-F^*),$$

where F is a strictly upper triangular $N\times N$ operator matrix whose entries are bounded linear operators on $L_2^m([a,b])$ and the off diagonal entries in the middle term of the right hand side of (4.2) are equal to the zero operator on $L_2^m([a,b])$.

PROPOSITION 4.1. *Assume that the operators K_{ij}, $i,j=1,\ldots,N$, in (4.1) are in the Perelson class. Then K and the diagonal entries $D_1,\ldots,D_N$ in (4.2) also belong to the Perelson class and*

$$(4.3)\qquad \mathrm{DET}(I-K)=\prod_{j=1}^{N}\mathrm{DET}(I-D_j).$$

PROOF. Let M_h be the averaging operator introduced in the previous section. Put $(K_{ij})_h=M_hK_{ij}M_h$ for each i and j. Furthermore, let $\widetilde{M_h}$ be the $N\times N$ diagonal operator matrix whose diagonal entries are equal to M_h, i.e.,

$$\widetilde{M_h}=\begin{pmatrix} M_h & & & \\ & M_h & & \\ & & \ddots & \\ & & & M_h \end{pmatrix}.$$

Note that $\widetilde{M_h}$ is the averaging operator on $L_2^{mN}([a,b])$. Now

$$\mathrm{tr}(\widetilde{M_h}K\widetilde{M_h})=\sum_{i=1}^{N}\mathrm{tr}(K_{ii})_h\to\sum_{i=1}^{N}\mathrm{TR}(K_{ii}),\qquad h\downarrow 0,$$

because of our hypotheses on $K_{11},\ldots,K_{NN}$. So we see that K belongs to the Perelson class and

$$(4.4)\qquad \mathrm{TR}(K)=\mathrm{TR}(K_{11})+\cdots+\mathrm{TR}(K_{NN}).$$

From [BarG] we know that the diagonal term in the right hand side of (4.2) is determined by the following identities:

$$(4.5a)\qquad D_i=K_{ii}+(K_{i,i+1}\ K_{i,i+2}\ \cdots\ K_{i,N})\times$$

$$\times\begin{pmatrix} I-K_{i+1,i+1} & -K_{i+1,i+2} & \cdots & -K_{i+1,N}\\ -K_{i+2,i+1} & I-K_{i+2,i+2} & \cdots & -K_{i+2,N}\\ \vdots & \vdots & & \vdots\\ -K_{N,i+1} & -K_{N,i+2} & \cdots & I-K_{N,N}\end{pmatrix}^{-1}\begin{pmatrix}K_{i+1,i}\\ K_{i+2,i}\\ \vdots\\ K_{N,i}\end{pmatrix},\qquad i=1,\ldots,N-1;$$

(4.5b) $$D_N = K_{NN}.$$

From these identities we see that $D_i - K_{ii}$ is a trace class operator for $i = 1, \dots, N$. So we can apply [GK1], Propositions II.1.1 and II.1.2 to show that $D_1, \dots, D_N$ are in the Perelson class.

From (4.2) it follows that the entries of F are Hilbert-Schmidt operators. Indeed, write $I - D$ for the middle term in the right hand side of (4.2), and put $H = I - (I - F^*)^{-1}$. Then H is strictly lower triangular, and (4.2) implies that $F - H = K - KH + FD - D$. Thus $F - H$ is a Hilbert-Schmidt operator. Since F is strictly upper triangular and H is strictly lower triangular, we may conclude that the entries of F are Hilbert-Schmidt.

Note that F is nilpotent, because F is strictly upper triangular. So F has no nonzero eigenvalue, and therefore $\widetilde{\det}(I - F) = 1$. Also FD is strictly upper triangular, and thus $\operatorname{tr} FD = 0$. By duality also $\widetilde{\det}(I - F^*) = 1$. Now, use the multiplication rule for the second regularized determinant (see formula (2.5) in Chapter IV of [GKr1]). It follows that

(4.6) $$\begin{aligned} \widetilde{\det}(I - K) &= \widetilde{\det}\big((I - F)(I - D)\big)e^{-\operatorname{tr} CF^*} \\ &= \widetilde{\det}(I - D)e^{-\operatorname{tr} CF^*}, \end{aligned}$$

where $C = F + D - FD$. From (4.2) we know that $K = C + F^* - CF^*$, and hence

$$CF^* = F + D - FD + F^* - K.$$

Recall that F and FD are strictly upper triangular and F^* is strictly lower triangular. So the diagonal entries of CF^* coincide with the diagonal entries of $D - K$. Therefore

(4.7) $$\begin{aligned} \operatorname{tr} CF^* &= \sum_{i=1}^{N} \operatorname{tr}(CF^*)_{ii} \\ &= \sum_{i=1}^{N} \operatorname{TR}(D_i) - \sum_{i=1}^{N} \operatorname{TR}(K_{ii}) \\ &= \operatorname{TR}(D) - \operatorname{TR}(K). \end{aligned}$$

For the third equality we apply (4.4) both to K and D.

By the result of the first paragraph of the proof (applied to both K and D) the operators K and D are in the Perelson class. But then we can use the second equality in (4.6) and the third in (4.7) to show that

(4.8) $$\operatorname{DET}(I - K) = \operatorname{DET}(I - D).$$

Since D is a diagonal operator matrix with diagonal entries in the Perelson class, it is clear from the definition of the Perelson determinant that the right hand side of (4.8) is equal to the right hand side of (4.3), and hence (4.3) is proved. □

II. MAIN RESULTS

II.1 Multiplicative diagonal and entropy integral

Let T be an element of the algebra $\mathcal{B}(\sigma)$ (see Section I.1 for the definition), and assume that $I-T$ is positive definite. Let $(T_{i-j}^{(\sigma)})_{i,j=-\infty}^{\infty}$ be the σ-block partitioning of T. We know that its symbol,

$$S_{T,\sigma}(z) = \sum_{\nu=-\infty}^{\infty} T_{\nu}^{(\sigma)} z^{\nu}$$

is in the Wiener algebra on $\mathbb{T}$ over the Hilbert-Schmidt operators on $L_2^m([a,b])$. Since $I-T$ is a positive definite operator, $I - S_{T,\sigma}(z)$ has (see Section I.2) a well-defined right multiplicative diagonal which we shall denote by $\Delta_0(I-T;\sigma)$. From Proposition I.2.1 we know that

$$\Delta_0(I-T;\sigma) = I - T_0^{(\sigma)} - B^{(\sigma)}(I-D^{(\sigma)})^{-1}C^{(\sigma)}, \tag{1.1}$$

where

$$D^{(\sigma)}\colon \oplus_1^{\infty} L_2^m([0,\sigma]) \to \oplus_1^{\infty} L_2^m([0,\sigma]),$$
$$C^{(\sigma)}\colon L_2^m([0,\sigma]) \to \oplus_1^{\infty} L_2^m([0,\sigma]),$$
$$B^{(\sigma)}\colon \oplus_1^{\infty} L_2^m([0,\sigma]) \to L_2^m([0,\sigma]),$$

are defined as follows:

$$\left(D^{(\sigma)}(\varphi_j)_{j=1}^{\infty}\right)_i = \sum_{j=1}^{\infty} T_{i-j}^{(\sigma)}\varphi_j, \qquad i = 1,2,\ldots,$$

$$C^{(\sigma)}\varphi = (T_i^{(\sigma)}\varphi)_{i=1}^{\infty}, \qquad B^{(\sigma)}(\varphi_j)_{j=1}^{\infty} = \sum_{j=1}^{\infty} T_{-j}^{(\sigma)}\varphi_j.$$

In particular,

$$\Delta_0(I-T;\sigma) - (I - T_0^{(\sigma)}) \in S_1, \tag{1.2}$$

where S_1 stands for the operator ideal of trace class operators.

Let f be an $m\times m$ matrix function with entries in $L_1(\mathbb{R})\cap L_2(\mathbb{R})$. The convolution operator on $L_2^m(\mathbb{R})$ associated with f will be denoted by L_f, i.e.,

$$(L_f\varphi)(t) = \int_{-\infty}^{\infty} f(t-s)\varphi(s)ds, \qquad t\in\mathbb{R}.$$

From Proposition I.4.1 in [GK2] we know that $L_f \in \mathcal{B}(\sigma)$ for each $\sigma > 0$, and thus $\Delta_0(I-L_f;\sigma)$ is a well-defined operator for each σ whenever $I-L_f$ is positive definite. The next theorem describe the entropy integral of f,

$$\mathcal{E}(f) = \lim_{\varepsilon\downarrow 0}\left(\frac{1}{2\pi}\int_{-\infty}^{\infty}\frac{\log\det(I-\widehat{f}(\lambda))}{\varepsilon^2\lambda^2+1}d\lambda\right), \tag{1.3}$$

in terms of $\Delta_0(I - L_f; \sigma)$.

THEOREM 1.1. *Let f be an $m \times m$ matrix function with entries in $L_1(\mathbb{R}) \cap L_2(\mathbb{R})$, and assume that there exists $\delta > 0$ such that f has continuous extensions defined on the closed intervals $[-\delta, 0]$ and $[0, \delta]$. Furthermore, assume that $I - L_f$ is positive definite. Then $\Delta_0(I - L_f; \sigma) - I$ is in the Perelson class and*

$$(1.4) \qquad \left(\operatorname{DET} \Delta_0(I - L_f; \sigma)\right)^{1/\sigma} = e^{\mathcal{E}(f)}.$$

In particular, the left hand side of (1.4) is independent of σ.

Theorem 1.1 is just a reformulation of Theorem I.5.2 in [GK1].

II.2 Approximation from the outside

Let k be an $m \times m$ matrix function with entries in $L_2([-\tau, \tau])$, and assume that k satisfies the following additional condition:

(C) there exists $\delta > 0$ such that k has continuous extensions defined on the closed intervals $[-\delta, 0]$ and $[0, \delta]$.

Furthermore, let f be an $m \times m$ matrix function with entries in $L_1(\mathbb{R}) \cap L_2(\mathbb{R})$, and let f be a positive extension of k, i.e.,

(a) $f(t) = k(t)$ for $-\tau \le t \le \tau$,

(b) $I - L_f$ is a positive definite operator.

In that case

$$(2.1) \qquad ((I - L_f)^{-1}\varphi)(t) = \varphi(t) - \int_{-\infty}^{\infty} \gamma(t - s)\varphi(s)ds, \qquad t \in \mathbb{R},$$

for a unique $m \times m$ matrix function γ. We shall assume that f is not equal to the band extension b of k, that is, the support of γ is not contained in $[-\tau, \tau]$. Our aim is to prove

$$(2.2) \qquad \mathcal{E}(f) < \mathcal{E}(b).$$

The proof of (2.2) will be based on two different approximations. The first is considered in the present section.

Put $\sigma_n = \frac{1}{n}\tau$, and let $(F_{i-j}^{(n)})_{i,j=-\infty}^{\infty}$ be the σ_n-block partitioning of L_f. The fact that f is an extension of k implies that the operators $F_j^{(n)}$, $|j| \le n - 1$, are uniquely determined by k. In fact,

$$(2.3) \qquad (F_j^{(n)}\varphi)(t) = \int_0^{\sigma_n} k(t - s + j\sigma_n)\varphi(s)ds, \qquad 0 \le t \le \sigma_n,\ |j| \le n - 1.$$

In this section we consider the band

$$(2.4) \qquad I - \sum_{j=-n}^{n} z^j F_j^{(n)}.$$

By $I - C^{(n)}(\cdot)$ we denote the band extension of the discrete operator band (2.4), and we write $\widehat{C}^{(n)}$ for the operator on $L_2^m(\mathbb{R})$ whose σ_n-block partitioning is the $L_2^m([0,\sigma_n])$-block Laurent operator with symbol $C^{(n)}(\cdot)$. According to Proposition II.1.1 in [GK2] the function $C^{(n)}(\cdot)$ is in the Wiener algebra on $\mathbb{T}$ over the Hilbert-Schmidt operators on $L_2^m([0,\sigma_n])$, and hence $\widehat{C}^{(n)} \in \mathcal{B}(\sigma_n)$.

Since $I - C^{(n)}(z)$ is a positive definite operator on $L_2^m([0,\sigma_n])$ for each $z \in \mathbb{T}$, the operator $I - \widehat{C}^{(n)}$ is positive definite, and thus $\Delta_0(I - \widehat{C}^{(n)}; \sigma_n)$ is a well-defined operator. By applying (1.1) and (1.2) with T replaced by $\widehat{C}^{(n)}$ we see that

(2.5) $$\Delta_0(I - \widehat{C}^{(n)}; \sigma_n) - (I - F_0^{(n)}) \in S_1.$$

Note that

(2.6) $$(F_0^{(n)}\varphi)(t) = \int_0^{\sigma_n} k(t-s)\varphi(s)ds, \qquad 0 \le t \le \sigma_n,$$

and thus $F_0^{(n)}$ belongs to the Perelson class $\mathcal{P}$ because of condition (C) on k. In fact, (see [GK1], Proposition II.1.4),

(2.7) $$\operatorname{TR}(F_0^{(n)}) = \frac{\tau}{2n} \operatorname{tr}\{k(0+) + k(0-)\}.$$

Since $F_0^{(n)}$ belongs to the Perelson class, formula (2.5) implies that the same holds true for $\Delta_0(I - \widehat{C}^{(n)}; \sigma_n) - I$.

THEOREM 2.1. *Let f be an $m \times m$ matrix function with entries in $L_1(\mathbb{R}) \cap L_2(\mathbb{R})$, and let f be a positive extension of k. Assume that f is not the band extension of k. Then there exist a positive integer n and $\eta > 0$ such that*

$$\left(\operatorname{DET} \Delta_0(I - \widehat{C}^{(pn)}; \sigma_{pn})\right)^{1/\sigma_{pn}} \ge \eta + \exp(\mathcal{E}(f)), \qquad p = 1, 2, \ldots .$$

To prove the above theorem we need the following lemma.

LEMMA 2.2. *For $p = 1, 2, 3, \ldots$ we have*

(2.8) $$\left(\operatorname{DET} \Delta_0(I - \widehat{C}^{(pn)}; \sigma_{pn})\right)^{pn} \ge \left(\operatorname{DET} \Delta_0(I - \widehat{C}^{(n)}; \sigma_n)\right)^n.$$

PROOF. Put $\ell = pn$, and consider the band

(2.9) $$I - \sum_{j=-\ell}^{\ell} z^j F_j^{(\ell)}.$$

We need the nonstationary operator Wiener algebra $\mathcal{NSW}(L_2^m([0,\sigma_\ell]))$. Note that the σ_ℓ-block partitioning of $I - \widehat{C}^{(n)}$ is an element of the latter algebra. In fact, the σ_ℓ-block partitioning of $I - \widehat{C}^{(n)}$ is a positive extension in $\mathcal{NSW}(L_2^m([0,\sigma_\ell]))$ of the band (2.9). Let

(2.10) $$\Delta(I - \widehat{C}^{(n)}; \sigma_\ell) = \operatorname{diag}\left(\Delta_i(I - \widehat{C}^{(n)}; \sigma_\ell)\right)_{i=-\infty}^{\infty}$$

be the right multiplicative diagonal of the σ_ℓ-block partitioning of $I-\widehat{C}^{(n)}$. Since $I-\widehat{C}^{(\ell)}$ is the band extension corresponding to the band (2.9), the maximum entropy principle for the band (2.9) in $\mathcal{NSW}(L_2^m([0,\sigma_\ell]))$ implies (cf., Section 2.4 in [GKW3]) that

$$\Delta_i(I-\widehat{C}^{(n)};\sigma_\ell) \leq \Delta_0(I-\widehat{C}^{(\ell)};\sigma_\ell). \tag{2.11}$$

Here we used that the diagonal entries of the right multiplicative diagonal of the σ_ℓ-block partitioning of $I-\widehat{C}^{(\ell)}$ do not depend on i and are all equal to $\Delta_0(I-\widehat{C}^{(\ell)};\sigma_\ell)$. From the construction of the diagonal entries of the right multiplicative diagonal (cf., formula (I.2.3)), it follows that

$$\Delta_i(I-\widehat{C}^{(n)};\sigma_\ell)-(I-F_0^{(\ell)}) \in S_1, \qquad i\in\mathbb{Z}.$$

Also,

$$\Delta_0(I-\widehat{C}^{(\ell)};\sigma_\ell)-(I-F_0^{(\ell)}) \in S_1.$$

Thus for each i

$$\Delta_i(I-\widehat{C}^{(n)};\sigma_\ell)-\Delta_0(I-\widehat{C}^{(\ell)};\sigma_\ell) \in S_1.$$

But then we can use the separation lemma for the Perelson determinant (see Section I.3) to conclude that

$$\mathrm{DET}\,\Delta_i(I-\widehat{C}^{(n)};\sigma_\ell) \leq \mathrm{DET}\,\Delta_0(I-\widehat{C}^{(\ell)};\sigma_\ell), \qquad i=0,\pm1,\pm2,\dots . \tag{2.12}$$

The next step is to compare $\Delta_0(I-\widehat{C}^{(n)};\sigma_n)$ with $\Delta_i(I-\widehat{C}^{(n)};\sigma_\ell)$ for $i=0,\pm1,\pm2,\dots$. Put

$$R := \Delta_0(I-\widehat{C}^{(n)};\sigma_n)-(I-F_0^{(n)}). \tag{2.13}$$

Both $\Delta_0(I-\widehat{C}^{(n)};\sigma_n)$ and R act on $L_2^m([0,\sigma_n])$. We know (see (2.5)) that R is a trace class operator. Let

$$\begin{pmatrix} \Delta_0^{(1,1)} & \cdots & \Delta_0^{(1,p)} \\ \vdots & & \vdots \\ \Delta_0^{(p,1)} & \cdots & \Delta_0^{(p,p)} \end{pmatrix}, \qquad \begin{pmatrix} R_{11} & \cdots & R_{1p} \\ \vdots & & \vdots \\ R_{p1} & \cdots & R_{pp} \end{pmatrix} \tag{2.14}$$

be the σ_ℓ-block partitioning of $\Delta_0(I-\widehat{C}^{(n)};\sigma_n)$ and R, respectively. Denote the first matrix by Δ_0. From (2.13) we see that

$$R_{ii}=\Delta_0^{(i,i)}-(I-F_0^{(\ell)}), \qquad i=1,\dots,p. \tag{2.15}$$

Since $F_0^{(\ell)}$ belongs to the Perelson class, (2.15) implies that the same holds true for $I-\Delta_0^{(i,i)}$, where $i=1,\dots,p$. From the result proved in the first paragraph of the proof of Proposition I.4.1 it follows that $I-\Delta_0$ belongs to the Perelson class and

$$\mathrm{TR}(I-\Delta_0)=\sum_{i=1}^{N}\mathrm{TR}(I-\Delta_0^{(i,i)}). \tag{2.16}$$

Now use (2.15), (2.7), (2.11) and the fact that R is a trace class operator to show that

$$\begin{aligned}\sum_{i=1}^{p} \operatorname{TR}(I-\Delta_0^{(i,i)}) &= \sum_{i=1}^{p} \operatorname{TR}(F_0^{(\ell)} - R_{ii}) \\ &= p\operatorname{TR}(F_0^{(\ell)}) - \sum_{i=1}^{p} \operatorname{tr}(R_{ii}) \\ &= \operatorname{TR}(F_0^{(n)}) - \operatorname{tr}(R) \\ &= \operatorname{TR}(F_0^{(n)} - R) \\ &= \operatorname{TR}\big(I - \Delta_0(I-\widehat{C}^{(n)};\sigma_n)\big).\end{aligned}$$

So we have $\operatorname{TR}\big(I-\Delta_0(I-\widehat{C}^{(n)};\sigma_n)\big) = \operatorname{TR}(I-\Delta_0)$. Since $\Delta_0(I-\widehat{C}^{(n)};\sigma_n)$ and Δ_0 are unitarily equivalent, these operators have the same second regularized determinant. Hence we have proved that

$$\operatorname{DET}\Delta_0(I-\widehat{C}^{(n)};\sigma_n) = \operatorname{DET}\Delta_0. \tag{2.17}$$

Since $\Delta_0(I-\widehat{C}^{(n)};\sigma_n)$ is positive definite, the same holds true for Δ_0, and so Δ_0 factors as

$$\Delta_0 = (I+A)\begin{pmatrix} D_1 & & \\ & \ddots & \\ & & D_p \end{pmatrix}(I+A^*), \tag{2.18}$$

where A is a strictly upper triangular $p\times p$ operator matrix and the off diagonal entries of the middle term in the right hand side of (2.18) are equal to zero. By Proposition I.4.1 (applied to $K = I-\Delta_0$) and (2.17) we have

$$\operatorname{DET}\Delta_0(I-\widehat{C}^{(n)};\sigma_n) = \prod_{i=1}^{p}\operatorname{DET}D_i. \tag{2.19}$$

Since the σ_n-block partitioning of $I-\widehat{C}^{(n)}$ is a $L_2^m([0,\sigma_n])$-block Laurent operator, the σ_ℓ-block partitioning of $I-\widehat{C}^{(n)}$ is p-periodic. From Section I.2 we know that

$$\Delta(I-\widehat{C}^{(n)};\sigma_\ell) = \operatorname{diag}(\ldots,D_1,D_2,\ldots,D_p,D_1,D_2,\ldots,D_p,\ldots).$$

So we have, by (2.12),

$$\operatorname{DET}D_i \le \operatorname{DET}\Delta_0(I-\widehat{C}^{(\ell)};\sigma_\ell), \qquad i=1,\ldots,p.$$

By combining this with (2.19) we obtain

$$\operatorname{DET}\Delta_0(I-\widehat{C}^{(n)};\sigma_n) \le \big(\operatorname{DET}\Delta_0(I-\widehat{C}^{(\ell)};\sigma_\ell)\big)^p.$$

Since $\ell = pn$, the lemma is proved. □

PROOF OF THEOREM 2.1. First we show that there exists a positive integer n such that

$$I - \widehat{C}^{(n)} \neq I - L_f. \tag{2.20}$$

The proof is by contradiction. Let γ be as in (2.1). Assume $I - \widehat{C}^{(n)} = I - L_f$. Recall that $I - C^n(\cdot)$, the symbol of the σ_n-block partitioning of $I - \widehat{C}^{(n)}$ is the band extension of the band (2.4). It follows that the j-th Fourier coefficient of $\left(I - C^{(n)}(\cdot)\right)^{-1}$ is zero for $|j| \geq n+1$. In other words,

$$\left(I - C^{(n)}(z)\right)^{-1} = I - \sum_{j=-n}^{n} z^j G_j^{(n)} \tag{2.21}$$

for certain $G_j^{(n)}$. Since $I - \widehat{C}^{(n)} = I - L_f$ by assumption, $\left(I - C^{(n)}(\cdot)\right)^{-1}$ is the symbol of the σ_n-block partitioning of $I - L_f$, and hence (2.21) implies that the set

$$\left\{(t,s) \mid \gamma(t-s) \neq 0,\ |t-s| \geq \tau + \frac{1}{n}\tau\right\}$$

must have measure zero. This holds for each n, and therefore

$$\{(t,s) \mid \gamma(t-s) \neq 0,\ |t-s| \geq \tau\}$$

has measure zero. But then $\operatorname{supp}\gamma \subset [-\tau, \tau]$, and thus f is the band extension of k, which contradicts our hypotheses on f. So for some n the inequality (2.20) holds true.

Since $I - L_f$ is positive definite, $I - \sum_{j=-\infty}^{\infty} z^j F_j^{(n)}$ is a positive extension of the band (2.4) which, by (2.20), is different from the band extension $I - C^{(n)}(\cdot)$. So we can apply the maximum entropy principle for discrete bands, to conclude that

$$\Delta_0(I - \widehat{C}^{(n)}; \sigma_n) > \Delta_0(I - L_f; \sigma_n).$$

Since

$$\Delta_0(I - \widehat{C}^{(n)}; \sigma_n) - (I - F_0^{(n)}) \in S_1,$$

$$\Delta_0(I - L_f; \sigma_n) - (I - F_0^{(n)}) \in S_1,$$

the difference $\Delta_0(I - \widehat{C}^{(n)}; \sigma_n) - \Delta_0(I - L_f; \sigma_n)$ is a trace class operator, and we can use Lemma I.3.1 to show that

$$\operatorname{DET}\Delta_0(I - \widehat{C}^{(n)}; \sigma_n) > \operatorname{DET}\Delta_0(I - L_f; \sigma_n).$$

From Theorem 1.1 we know that

$$\left(\operatorname{DET}\Delta_0(I - L_f; \sigma_n)\right)^{1/\sigma_n} = \exp(\mathcal{E}(f)),$$

and so we have shown that

$$\operatorname{DET}\Delta_0(I-\widehat{C}^{(n)};\sigma_n)^{1/\sigma_n} > \exp(\mathcal{E}(f)).$$

Put

$$\eta := \operatorname{DET}\Delta_0(I-\widehat{C}^{(n)};\sigma_n)^{1/\sigma_n} - \exp(\mathcal{E}(f)). \tag{2.22}$$

Since $\sigma_n = \frac{1}{n}\tau$ and $\sigma_{pn} = \frac{1}{pn}\tau$, the proof is completed by taking η as in (2.22) and by applying Lemma 2.2. □

II.3 Approximation from the inside

Let k and f be as in the first paragraph of Section II.2. Again, put $\sigma_n = \frac{1}{n}\tau$, and let $(F_{i-j}^{(n)})_{i,j=-\infty}^{\infty}$ be the σ_n-block partitioning of L_f. In this section we consider the band

$$I - \sum_{j=-n+1}^{n-1} z^j F_j^{(n)}. \tag{3.1}$$

From (2.3) we know that the coefficients $F_j^{(n)}$ in (3.1) are uniquely determined by k and do not depend on the choice of f.

By $I - B^{(n)}(\cdot)$ we denote the band extension of the discrete operator band (3.1). We write $\widehat{B}^{(n)}$ for the operator on $L_2^m(\mathbb{R})$ whose σ_n-block partitioning is the $L_2^m([0,\sigma_n])$-block Laurent operator with symbol $B^{(n)}(\cdot)$. According to Proposition II.1.1 in [GK2] the function $B^{(n)}(\cdot)$ is in the Wiener algebra on $\mathbb{T}$ over the Hilbert-Schmidt operators on $L_2^m([0,\sigma_n])$, and hence $\widehat{B}^{(n)} \in \mathcal{B}(\sigma_n)$.

Since $I - B^{(n)}(z)$ is a positive definite operator on $L_2^m([0,\sigma_n])$ for each $z \in \mathbb{T}$, the operator $I - \widehat{B}^{(n)}$ is positive definite, and thus $\Delta_0(I-\widehat{B}^{(n)};\sigma_n)$ is a well-defined operator. By applying (1.1) and (1.2) with $\widehat{B}^{(n)}$ in place of T we see that

$$\Delta_0(I-\widehat{B}^{(n)};\sigma_n) - (I - F_0^{(n)}) \in S_1. \tag{3.2}$$

Since $F_0^{(n)}$ belongs to the Perelson class (see the previous section), formula (3.2) implies that the same holds true for $\Delta_0(I-\widehat{B}^{(n)};\sigma_n) - I$.

THEOREM 3.1. *We have*

$$\lim_{n\to\infty}\left(\operatorname{DET}\Delta_0(I-\widehat{B}^{(n)};\sigma_n)\right)^{1/\sigma_n} = e^{\mathcal{E}(b)}, \tag{3.3}$$

where b is the band extension of k.

Since $\sigma_n = \frac{1}{n}\tau$, we know from Proposition I.5.2 in [GK2] that $\widehat{B}^{(n)}$ belongs to $\mathcal{B}(\tau)$. Hence we may consider the right multiplicative diagonal $\Delta_0(I-\widehat{B}^{(n)};\tau)$ of the symbol of the τ-block partitioning of $I-\widehat{B}^{(n)}$. We shall need the following lemma.

LEMMA 3.2. *For $n = 1, 2, \ldots$ we have*

$$(3.4)\qquad \left(\operatorname{DET}\Delta_0(I - \widehat{B}^{(n)};\sigma_n)\right)^{1/\sigma_n} = \left(\operatorname{DET}\Delta_0(I - \widehat{B}^{(n)};\tau)\right)^{1/\tau}.$$

PROOF. Let K be the operator on $L_2^m([0,\tau])$ defined by

$$(3.5)\qquad (K\varphi)(t) = \int_0^\tau k(t-s)\varphi(s)ds, \qquad 0 \le t \le \tau.$$

We know that for each $n \ge 1$

$$(3.6)\qquad \Delta_0(I - \widehat{B}^{(n)};\tau) - (I - K) \in S_1.$$

Fix $n \ge 1$. Let

$$(3.7)\qquad \begin{pmatrix} \Delta_{11} & \cdots & \Delta_{1n} \\ \vdots & & \vdots \\ \Delta_{n1} & \cdots & \Delta_{nn} \end{pmatrix}$$

be the σ_n-block partitioning of $\Delta_0(I - \widehat{B}^{(n)};\tau)$, and consider the factorization:

$$\begin{pmatrix} \Delta_{11} & \cdots & \Delta_{1n} \\ \vdots & & \vdots \\ \Delta_{n1} & \cdots & \Delta_{nn} \end{pmatrix} = (I + A)\begin{pmatrix} D_1 & & 0 \\ & \ddots & \\ 0 & & D_n \end{pmatrix}(I + A^*),$$

where A is a strictly upper triangular $n \times n$ operator matrix. Since the σ_n-block partitioning of $I - \widehat{B}^{(n)}$ is a $L_2^m([0,\sigma_n])$-block Laurent operator, we have

$$D_1 = \cdots = D_n = \Delta_0(I - \widehat{B}^{(n)};\sigma_n).$$

Let T be the trace class operator defined by the left hand side of (3.6), and let

$$\begin{pmatrix} T_{11} & \cdots & T_{1n} \\ \vdots & & \vdots \\ T_{n1} & \cdots & T_{nn} \end{pmatrix}$$

be the σ_n-block partitioning of T. Then

$$(3.8)\qquad \Delta_{ii} = T_{ii} + (I - F_0^{(n)}), \qquad i = 1, \ldots, n.$$

Hence $I - \Delta_{11}, \ldots, I - \Delta_{nn}$ are in the Perelson class, and we can apply Proposition I.4.1 to show that

$$\operatorname{DET}\Delta = \prod_{i=1}^n \operatorname{DET} D_i,$$

where Δ is the $n \times n$ operator matrix in (3.7). Furthermore, by the result of the first paragraph of the proof of Proposition I.4.1,

$$\operatorname{TR}(I - \Delta) = \sum_{i=1}^n \operatorname{TR}(I - \Delta_{ii}).$$

On the other hand, by (3.8), (2.7) and (3.6),

$$\begin{aligned}\sum_{i=1}^{n}\operatorname{TR}(I-\Delta_{ii}) &= \sum_{i=1}^{n}\operatorname{TR}(F_0^{(n)}-T_{ii})\\ &= \sum_{i=1}^{n}\Big[\frac{\tau}{2n}\operatorname{tr}\{k(0+)-k(0-)\}-\operatorname{tr}T_{ii}\Big]\\ &= \frac{\tau}{2}\operatorname{tr}\{k(0+)-k(0-)\}-\sum_{i=1}^{n}\operatorname{tr}T_{ii}\\ &= \operatorname{TR}(K)-\operatorname{tr}T\\ &= \operatorname{TR}(I-\Delta_0(I-\widehat{B}^{(n)};\tau)).\end{aligned}$$

Since Δ and $\Delta_0(I-\widehat{B}^{(n)};\tau)$ are unitarily equivalent, we conclude that

$$\begin{aligned}\operatorname{DET}\Delta_0(I-\widehat{B}^{(n)};\tau) = \operatorname{DET}\Delta &= \prod_{i=1}^{n}\operatorname{DET}D_i\\ &= \big(\operatorname{DET}\Delta_0(I-\widehat{B}^{(n)};\sigma_n)\big)^n.\end{aligned}$$

Since $\sigma_n = \frac{1}{n}\tau$, this yields (3.4). □

PROOF OF THEOREM 3.1. From [GK2], Theorem III.1.1 we know that

(3.9) $$\|\widehat{B}^{(n)}-L_b\|_{\mathcal{B}(\tau)}\to 0 \qquad (n\to\infty).$$

Let $(\widehat{B}_{i-j}^{(n)})_{i,j=-\infty}^{\infty}$ be the τ-block partitioning of $\widehat{B}^{(n)}$, and let $(\widehat{B}_{i-j})_{i,j=-\infty}^{\infty}$ be the τ-block partitioning of L_b. Then formula (3.9) tells us that

(3.10a) $$\sum_{\nu=-\infty}^{\infty}\|\widehat{B}_\nu^{(n)}-\widehat{B}_\nu\|\to 0 \qquad (n\to\infty),$$

(3.10b) $$\sum_{\nu=-\infty}^{\infty}\|\widehat{B}_\nu^{(n)}-\widehat{B}_\nu\|_2^2\to 0 \qquad (n\to\infty).$$

Note that $\widehat{B}_0^{(n)}$ is precisely the operator K defined by (3.5). It follows that

$$\Delta_0(I-\widehat{B}^{(n)};\tau) = I-K+E^{(n)}(I-T^{(n)})F^{(n)},$$

where

$$F^{(n)}: L_2^m([0,\tau]) \to \oplus_1^\infty L_2^m([0,\tau]), \qquad F^{(n)} = \begin{pmatrix} \widehat{B}_1^{(n)} \\ \widehat{B}_2^{(n)} \\ \vdots \end{pmatrix};$$

$$E^{(n)}: \oplus_1^\infty L_2^m([0,\tau]) \to L_2^m([0,\tau]), \qquad E^{(n)} = (\widehat{B}_{-1}^{(n)} \quad \widehat{B}_2^{(n)} \quad \cdots)$$

$$T^{(n)}: \oplus_1^\infty L_2^m([0,\tau]] \to L_2^m([0,\tau]), \qquad T^{(n)} = \begin{pmatrix} \widehat{B}_0^{(n)} & \widehat{B}_{-1}^{(n)} & \widehat{B}_{-2}^{(n)} & \cdots \\ \widehat{B}_1^{(n)} & \widehat{B}_0^{(n)} & \widehat{B}_{-1}^{(n)} & \cdots \\ \widehat{B}_2^{(n)} & \widehat{B}_1^{(n)} & \widehat{B}_0^{(n)} & \cdots \\ \vdots & \vdots & \vdots & \ddots \end{pmatrix}.$$

Next we consider $\Delta_0(I - L_b; \tau)$, the right multiplicative diagonal of the symbol of the τ-block partitioning of $I - L_b$. Note that $\widehat{B}_0$ is precisely the operator K defined by (3.7). Thus

$$\Delta_0(I - L_b; \tau) = I - K + E(I - T)F,$$

where E, F and T are the operators which one obtains if in the definitions of $E^{(n)}$, $F^{(n)}$ and $T^{(n)}$ the superscript (n) is omitted. From the limit in (3.10a) we see that

$$\|T^{(n)} - T\| \to 0 \qquad (n \to \infty).$$

The limit in (3.10b) implies that $E^{(n)} \to E$ and $F^{(n)} \to F$ in the Hilbert-Schmidt norm if $n \to \infty$. So we see that

$$\lim_{n\to\infty} E^{(n)}(I - T^{(n)})F^{(n)} = E(I - T)F \tag{3.11}$$

in the trace class norm as well as in the Hilbert-Schmidt norm. In particular, we have (cf., [GKr1], Theorem IV.2.1) that

$$\widetilde{\det}\big(I - K + E^{(n)}(I - T^{(n)})F^{(n)}\big) \to \widetilde{\det}\big(I - K + E(I - T)F\big), \qquad n \to \infty,$$

$$\operatorname{tr}\big(E^{(n)}(I - T^{(n)})F^{(n)}\big) \to \operatorname{tr} E(I - T)F, \qquad n \to \infty.$$

But then we may conclude that

$$\lim_{n\to\infty} \operatorname{DET} \Delta_0(I - \widehat{B}^{(n)}; \tau) = \operatorname{DET} \Delta_0(I - L_b; \tau).$$

By Theorem 1.1,

$$\big(\operatorname{DET} \Delta_0(I - L_b; \tau)\big)^{1/\tau} = \exp \mathcal{E}(b),$$

and therefore

$$\lim_{n\to\infty} \operatorname{DET} \Delta_0(I - \widehat{B}^{(n)}; \tau)^{1/\tau} = \exp \mathcal{E}(b).$$

Finally, one uses Lemma 3.2 to get the desired limit in (3.21). □

II.4 Proof of the maximum entropy principle

Let k and f be as in the first paragraph of Section II.2, and let b be the band extension of k. We assume that $f \neq b$, and we want to show that

$$\mathcal{E}(f) < \mathcal{E}(b). \tag{4.1}$$

Put $\sigma_n = \frac{1}{n}\tau$. Consider the operators $\widehat{C}^{(n)}$ and $\widehat{B}^{(n)}$ introduced in Sections II.2 and II.3, respectively. By definition, the σ_n-block partitioning of $I - \widehat{C}^{(n)}$ (resp., $I - \widehat{B}^{(n)}$) is the $L_2^m([0, \sigma_n])$-block Laurent operator whose symbol $I - C^{(n)}(\cdot)$ (resp., $I - B^{(n)}(\cdot)$) is the band extension of the discrete operator band (2.4) (resp., (3.1)). Since (3.1) is a part of (2.4), it follows that $I - C^{(n)}(\cdot)$ is also a positive extension of the band (3.1), and we can apply the maximum entropy principle for discrete bands (see [GKW3]) to show that

$$\Delta_0(I - \widehat{C}^{(n)}; \sigma_n) \leq \Delta_0(I - \widehat{B}^{(n)}; \sigma_n).$$

Now, by (2.5) and (3.2),

$$\Delta_0(I - \widehat{C}^{(n)}; \sigma_n) - \Delta_0(I - \widehat{B}^{(n)}; \sigma_n) \in S_1,$$

and thus (by Lemma I.3.1) we have

$$\big(\mathrm{DET}\, \Delta_0(I - \widehat{C}^{(n)}; \sigma_n)\big)^{1/\sigma_n} \leq \big(\mathrm{DET}\, \Delta_0(I - \widehat{B}^{(n)}; \sigma_n)\big)^{1/\sigma_n}, \qquad n \geq 1.$$

So far n was arbitrary. Now take n and $\eta > 0$ as in Theorem II.2.1. Then for $p = 1, 2, \ldots$

$$\big(\mathrm{DET}\, \Delta_0(I - \widehat{B}^{(pn)}; \sigma_{pn})\big)^{1/\sigma_{pn}} \geq \eta + \exp\big(\mathcal{E}(f)\big).$$

On the other hand, by Theorem II.3.1, the left hand side of the preceding inequality converges to $\exp\big(\mathcal{E}(b)\big)$ if $p \to \infty$. Thus

$$\exp\big(\mathcal{E}(b)\big) \geq \eta + \exp\big(\mathcal{E}(f)\big),$$

and (4.1) is proved.

REFERENCES

[AK] D.Z. Arov and M.G. Krein, Problems of search of the minimum of entropy in indeterminate extension problems, *Funct. Anal. Appl.* **15** (1981), 123–126.

[BarG] M.A. Barkar and I.C. Gohberg, On factorizations of operators relative to a discrete chain of projectors in Banach spaces, *Amer. Math. Soc. Transl.* (2) **90** (1970), 81–103. [*Mat. Issled.* **1** (1966), 32–54, Russian.]

[BGK] H. Bart, I. Gohberg and M.A. Kaashoek, *Minimal factorization of matrix and operator functions*, OT 1, Birkhäuser Verlag, Basel, 1979.

[Be] J.J. Benedetto, A quantitative maximum entropy theorem for the real line, *Integral Equations and Operator Theory* **10** (1987), 761–779.

[Ch] J. Chover, On normalized entropy and the extensions of a positive definite function, *J. Math. Mech.* **10** (1961), 927–945.

[D] H. Dym, *J Contractive matrix functions, reproducing kernel Hilbert spaces and interpolation*, CBMS 71, Amer. Math. Soc., Providence, RI, 1989.

[DG] H. Dym and I. Gohberg, On an extension problem, generalized Fourier analysis and an entropy formula, *Integral Equations and Operator Theory* **3** (1980), 143–215.

[GGK] I. Gohberg, S. Goldberg and M.A. Kaashoek, *Classes of Linear Operators*, Volume I, Birkhäuser Verlag, Basel, 1990.

[GK1] I. Gohberg and M.A. Kaashoek, Asymptotic formulas of Szegö-Kac-Achiezer type, *Asymptotic Analysis*, to appear.

[GK2] I. Gohberg and M.A. Kaashoek, The band extension of the real line as a limit of discrete band extensions, I. The main limit theorem, Proceedings Workshop on Operator Theory and Complex Analysis, Sapporo, Japan, 1991, to appear.

[GKW1] I. Gohberg, M.A. Kaashoek and H.J. Woerdeman, The band method for positive and contractive extension problems, *J. Operator Theory* **22** (1989), 109–155.

[GKW2] I. Gohberg, M.A. Kaashoek and H.J. Woerdeman, The band method for positive and contractive extension problems: An alternative version and new applications, *Integral Equations and Operator Theory* **12** (1989), 343–382.

[GKW3] I. Gohberg, M.A. Kaashoek and H.J. Woerdeman, A maximum entropy principle in the general framework of the band method, *J. Funct. Anal.* **95** (1991), 231–254.

[GKr1] I.C. Gohberg and M.G. Krein, *Introduction to the theory of nonselfadjoint operators*, Transl. Math. Monographs 18, Amer. Math. Soc., Providence, RI, 1969.

[LS] L.A. Ljusternik and W.I. Sobolev, *Elemente der Funktionalaysis*, Akademie-Verlag, Berlin, 1960.

[MG] D. Mustafa and K. Glover, *Minimum entropy H_∞ control*, Lecture Notes in Control and Information Sciences 146, Springer Verlag, Berlin, 1990.

[P] A. Perelson, Generalized traces and determinants for compact operators, Ph.D. Thesis, Tel-Aviv University, 1987.

I. Gohberg
School of Mathematical Sciences
The Raymond and Beverly Sackler Faculty of Exact Sciences
Tel-Aviv University
Tel-Aviv, Ramat-Aviv 69989, Israel

M.A. Kaashoek
Faculteit Wiskunde en Informatica
Vrije Universiteit
De Boelelaan 1081a
1081 HV Amsterdam, The Netherlands

Operator Theory:
Advances and Applications, Vol. 58

WEAKLY POSITIVE MATRIX MEASURES, GENERALIZED TOEPLITZ FORMS, AND THEIR APPLICATIONS TO HANKEL AND HILBERT TRANSFORM OPERATORS

Mischa Cotlar and Cora Sadosky[1]

The generalized Bochner theorem (GBT), that provides both integral representations for generalized Toeplitz forms in terms of positive matrix measures, and positive extensions of weakly positive ones, is the central result of a theory with many applications to Hankel and Hilbert transform operators.

The GBT unifies the proofs of the interpolation theorems of Nehari, Adamjan-Arov-Krein and Sarason, the lifting theorem of Sz. Nagy-Foias, and the prediction theorem of Helson-Szegö, and, in its more general abstract form, leads to multidimensional versions of these results.

The theory is presented here as related to the matrix extension problems solved by Gohberg and collaborators, as well as with complete positivity in C*-algebras.

INTRODUCTION

The study of weakly positive matrix measures and generalized Toeplitz kernels and forms started as an attempt to apply moment theory methods to the Hilbert transform [CS1,2]. In particular, it was shown in [CS2,3] that the Helson-Szegö theorem on the boundedness of the Hilbert transform in $L^2(\mu)$ spaces, can be deduced from a generalized Bochner theorem (GBT), which solves the following moment problem.

Given a scalar product $< \, , \, >$ in the space P of trigonometric polynomials, when does there exist a 2×2 matrix measure $[\mu_{ij}]$, $\mu_{ij} \geq 0$, $i,j = 1,2$, such that $\langle f,g\rangle = \Sigma_{i,j} \int f_i \overline{g_j} \, d\mu_{ij}$, for f_1, f_2, the analytic and

[1]Author partially supported by grants from the N.S.F. (USA).

antianalytic parts of f? The answer to this question lead to the notions in the title, and a theory of generalized Toeplitz forms has been developed in [CS1 - 19], [ArCS], [Ar1,2], [ArC], [Al1,2], [B], [DO1,2], [MO1,2], (cf. also the bibliography of [CS11]).

This paper is an expanded version of two talks on generalized Toeplitz kernels and forms, delivered by the authors at the Workshop on Continuous and Discrete Fourier Transforms and Extension Problems, held at the University of Maryland in September, 1991. The basic facts and some of the recent results in the theory are surveyed, concentrating mostly on the authors' contributions, and on aspects related to the main subject of this volume.

The notions of generalized Toeplitz (GT) form and weakly positive matrix measure are introduced in Section 1, where their lifting properties are obtained through the generalized Toeplitz theorem (GBT, Theorem 1), and some of their applications summarized. In Section 2, the notion of positivity in the set $M_2(P)$ of 2×2 matrices with polynomial entries is systematically explored, to obtain, from the GBT, the Helson-Szegö theorem for the Hilbert transform, as well as the Nehari theorem for Hankel operators. In Section 3, the abstract lifting theorem for abstract Hankel and essentially Hankel forms is derived from the Gelfand-Naimark-Segal (GNS) construction and the Wold-Kolmogorov decomposition. The lifting and n-conditional lifting theorems for forms acting in multiparametric scattering structures are stated in Section 4, where some examples on big and middle Hankel operators in $L^2(T^2)$ are given.

The relations between the work presented here and that of some other authors remain to be clarified. In particular, the relation between the order properties of $M_2(P)$ in Section 2 with the notion of complete positivity in C*-algebras (cf. [E] and [Arv]), as well as the relation between the theory of Hankel forms in Fock spaces provided by the GBT, and that developed by Janson, Peetre and Rochberg [cf. [JPR]). Although the relation of some of the questions discussed here with the band extension and four blocks problems treated by Gohberg and his collaborators is becoming more clear (cf. Section 1 and other papers in this volume), the connection should be further explored in order to improve results through "cross-fertilization."

1. LIFTING PROPERTIES OF GENERALIZED TOEPLITZ FORMS AND WEAKLY POSITIVE MATRIX MEASURES.

Here P denotes the vector space of all trigonometric polynomials spanned by the functions $e_n(t) = \exp(int)$, $n \in \mathbb{Z}$, and P_1, P_2 are the subspaces of the analytic and antianalytic polynomials, spanned by $\{e_n, n \geq 0\}$ and $\{e_n, n < 0\}$, respectively. Also, $P^2 = P \dot{+} P$ stands for the space of pairs $F = (F_1, f_2)$, $F_1, F_2 \in P$, and σ for the shift operator defined in P by

$$\sigma f(t) = \exp(it)f(t), \tag{1.1}$$

and in P^2 by

$$\sigma F = (\sigma f_1, \sigma f_2). \tag{1.1a}$$

$\mathbf{M}(\mathbf{T})$ is the set of the finite complex measures ρ defined in $\mathbf{T}$, $\mathbf{M}^+(\mathbf{T}) = \{\rho \in \mathbf{M}(\mathbf{T}): \rho \geq 0\}$, and $\mathbf{M}_2(\mathbf{T})$ denotes the set of 2×2 matrix measures, $\mu = [\mu_{ij}]$, where, for $i,j = 1,2$, $\mu_{ij} \in \mathbf{M}(\mathbf{T})$. Correspondingly, writing $\mu \geq 0$ when $[\mu_{ij}(\Delta)]$ is a non-negative definite matrix for each Borel set Δ in $\mathbf{T}$,

$$\mathbf{M}_2^+(\mathbf{T}) = \{\mu = [\mu_{ij}] \in \mathbf{M}_2(\mathbf{T}): \quad \mu \geq 0\} \tag{1.2}$$

The Generalized Bochner Theorem (GBT) (see Theorem 1 below) establishes a relation between sesquilinear forms (i.e., not necessarily positive scalar products) $\langle\ ,\ \rangle$ in P, and sesquilinear forms $\langle\langle\ ,\ \rangle\rangle$ in P^2, which leads to integral representations of the forms in P.

A form $\langle\ ,\ \rangle$ in P is said <u>positive</u> if $\langle f,f\rangle \geq 0$ $\forall f \in P$, and <u>Toeplitz</u> if

$$\langle f,g\rangle = \langle \sigma f, \sigma g\rangle, \ \forall f,g \in P. \tag{1.3}$$

Similarly, a forms $\langle\langle\ ,\ \rangle\rangle$ is positive in P^2 if $\langle\langle F,F \rangle\rangle \geq 0$, $\forall F \in P^2$, and Toeplitz if

$$\langle\langle F,G \rangle\rangle = \langle\langle \sigma F, \sigma G \rangle\rangle, \ \forall F,G \in P^2. \tag{1.3a}$$

Every $\rho \in \mathbf{M}(\mathbf{T})$ defines a Toeplitz form $\langle\ ,\ \rangle_\rho$, by

$$\langle f,g\rangle_\rho := \rho(f,g) = \int f\bar{g}d\rho \tag{1.4}$$

and $\langle\ ,\ \rangle_\rho$ is positive iff $\rho \geq 0$. The Herglotz-Boch theorem asserts that, conversely, a form $\langle\ ,\ \rangle$ in P is Toeplitz and positive iff $\langle\ ,\ \rangle = \langle\ ,\ \rangle_\rho$ for some $\rho \in \mathbf{M}^+(\mathbf{T})$, and this ρ is unique.

Similarly, every $\mu = [\mu_{ij}] \in \mathbf{M}_2(\mathbf{T})$ defines a Toeplitz form $\langle\langle\ ,\ \rangle\rangle_\mu$ in P^2, by

$$\langle\langle F,G \rangle\rangle_\mu = \langle\langle (f_1,f_2), (g_1,g_2) \rangle\rangle_\mu$$

$$:= \sum_{i,j=1,2} \mu_{ij}(f_i\bar{g}_j) = \sum_{i,j=1,2} \int f_i\bar{g}_j d\mu_{ij} \tag{1.5}$$

and $\langle\langle\ ,\ \rangle\rangle_\mu$ is positive iff $\mu \geq 0$ (see Section 2). In turn, the vector version of the Herglotz-Bochner theorem asserts that a form $\langle\langle\ ,\ \rangle\rangle$ in P^2 is Toeplitz and positive iff $\langle\langle\ ,\ \rangle\rangle = \langle\langle\ ,\ \rangle\rangle_\mu$ for some (unique) $\mu \in \mathbf{M}_2^+(\mathbf{T})$.

REMARK 1. If $\mu \in \mathbf{M}_2^+(\mathbf{T})$ and $\rho = \mu_{11} + \mu_{22}$, then $\rho \geq 0$ and $\mu_{ij} = \omega_{ij}(t)d\rho$ (see Section 2), so that

$$\langle\langle F,G\rangle\rangle_\mu = \int G^*(t)\ [\omega_{ij}(t)]\ F(t)d\rho(t) \tag{1.6}$$

for all $F,G \in P^2$.

Now, every $f \in P$ has a unique representation $f = P_1f + P_2f$, $P_1f \in P_1$, $P_2f \in P_2$, so that $f \to \pi f = (P_1f,\ P_2f)$ is a linear injective map of P into P^2, and P can be identified with the subspace πP of P^2. Through this map, every form $\langle\langle\ ,\ \rangle\rangle$ in P^2 induces an associated form $\langle\ ,\ \rangle$ in P, defined by

$$\langle f,g\rangle := \langle\langle f,\pi g\rangle\rangle,\ \forall f,g \in P. \tag{1.7}$$

Observe that if $\langle\langle\ ,\ \rangle\rangle$ is Toeplitz in P^2, this $\langle\ ,\ \rangle$ need not be Toeplitz in P, and only satisfies the weaker condition

$$\langle f,g\rangle = \langle \sigma f,\sigma g\rangle \text{ provided } \pi(\sigma f) = \sigma(\pi f),\ \pi(\sigma g) = \sigma(\pi g). \tag{1.8}$$

This leads to the following concept.

A form $\langle\ ,\ \rangle$ in P is called generalized Toeplitz (GT) if it satisfies the weaker condition (1.8), i.e, if for each i,j = 1,2, the restriction of the form to $P_i \times P_j$ is Toeplitz in the sense that, if both pairs (f,g) and $(\sigma f,\sigma g)$ belong to $P_i \times P_j$ then $\langle f,g\rangle = \langle \sigma f,\sigma g\rangle$, i,j = 1,2.

On the other hand, the associated form $\langle\ ,\ \rangle$ in P can be positive without $\langle\langle\ ,\ \rangle\rangle$ being positive in P^2, and, in such case, $\langle\langle\ ,\ \rangle\rangle$ is called weakly positive (WP).

In particular, every $\mu = [\mu_{ij}] \in \mathbf{M}_2(\mathbf{T})$ defines a Toeplitz form $\langle\langle\ ,\ \rangle\rangle_\mu$ in P^2, and a corresponding associated form $\langle\ ,\ \rangle_\mu$ in P,

$$\langle f,g\rangle_\mu := \sum_{i,j=1,2} \mu_{ij}((P_i f)(\overline{P_j g})) = \int G^*(t)[\omega_{ij}(t)]F(t)d\rho(t) \tag{1.9}$$

where $F = (P_1 f,\ P_2 f)$, $G = (P_1 g,\ P_2 g)$, and ρ is as in (1.6). Thus,

$$\langle f,g\rangle_\mu = \mu_{ij}(f\bar{g}) \quad \text{whenever} \quad f \in P_i,\ g \in P_j,\ i,j = 1,2. \tag{1.9a}$$

The form $\langle\ ,\ \rangle_\mu$ defined by (1.9) is a GT form and, if it is also positive, the measure μ is called weakly positive (WP) and denoted by $\mu \succ 0$. Thus, if $\mu \in \mathbf{M}_2(\mathbf{T})$, then

$$\mu \geq 0 \Leftrightarrow \langle\langle F,F \rangle\rangle_\mu = \sum_{i,j=1,2} \mu_{ij}(f_i\bar{f}_j) \geq 0,\ \forall F = (f_1,f_2) \in P^2, \tag{1.10}$$

while

$$\mu \succ 0 \Leftrightarrow \langle\langle F,F \rangle\rangle_\mu = \sum_{i,j=1,2} \mu_{ij}(f_i\bar{f}_j) \geq 0$$

$$\text{only if } F = (f_1,f_2) \in P_1 \times P_2 \subset P^2. \tag{1.10a}$$

Let us remark that scalar products $\langle\!\langle\ ,\ \rangle\!\rangle_\mu$, for $d\rho(t) = dt$ in (1.6), appear also in the work of Dym, Gohberg and their collaborators (cf. [EℓG] and other papers in this volume).

Our subject is based on the extension of scalar products in P to positive Toeplitz scalar products in P^2.

The basic result is the Generalized Bochner theorem (GBT):

THEOREM 1. [CS3]. *For a sesquilinear form* $\langle\ ,\ \rangle$ *in* P *the following conditions are equivalent:*

(1) $\langle\ ,\ \rangle$ *is a positive Toeplitz form.*

(2) *There exists a positive matrix measure* $\mu = [\mu_{ij}] \in \mathbf{M}_2^+(\mathbf{T})$ *such that* $\langle\ ,\ \rangle = \langle\ ,\ \rangle_\mu$, i.e.: $\langle f,g\rangle = \mu_{ij}(f\bar{g})$ *whenever* $f \in P_i$, $g \in P_j$, *for* $i,j = 1,2$.

(3) *There exists a positive Toeplitz form* $\langle\!\langle\ ,\ \rangle\!\rangle$ *in* P^2 *such that* $\langle f,g\rangle = \langle\!\langle \pi f, \pi g \rangle\!\rangle$, $\forall f,g \in P$ *and* π *as in* (1.7).

(4) $\langle\ ,\ \rangle$ *is positive and* $\langle\ ,\ \rangle = \langle\ ,\ \rangle_\mu$ *for some* $\mu \in \mathbf{M}_2(\mathbf{T})$.

For Toeplitz forms in P, the GBT reduces to the Herglotz-Bochner theorem.

COROLLARY 1. [CS3] (a) (Lifting property of GT forms.) *If* $\langle\ ,\ \rangle$ *is a positive GT form in* P, *then, by identifying* P *with* πP, $\langle\ ,\ \rangle$ *extends to a positive Toeplitz form* $\langle\!\langle\ ,\ \rangle\!\rangle$ *in* P^2, *i.e.:* $\langle f,g\rangle = \langle\!\langle \pi f, \pi g \rangle\!\rangle$.

(b) (Lifting property of weakly positive matrix measures.) *If* $\nu = [\nu_{ij}] \in \mathbf{M}_2(\mathbf{T})$ *is weakly positive,* $\nu \succ 0$, *then there exists a positive matrix measure*

$\mu = [\mu_{ij}] \in \mathbf{M}_2^+(\mathbf{T})$ *such that* $\langle\ ,\ \rangle_\nu = \langle\ ,\ \rangle_\mu$ *and, consequently,*

$$\mu_{11} = \nu_{11} \geq 0,\ \mu_{22} = \nu_{22} \geq 0,\ \mu_{12} = \nu_{12} + h\,dt,\ \mu_{21} = \nu_{21} + \bar{h}\,dt \qquad (1.11)$$

for $h \in H^1(\mathbf{T})$.

REMARK 2. Since there is a one-to-one correspondence between sesquilinear forms in P and kernels K: $\mathbb{Z} \times \mathbb{Z} \to \mathbb{C}$, given by $K(m,n) = \langle e_m, e_n \rangle$, the notions above can be given in terms of kernels, instead of forms. For instance, $\langle\ ,\ \rangle$ is positive iff K is positive definite, $\langle\ ,\ \rangle$ is Toeplitz iff $K(m + 1, n + 1) = K(m,n)$, $\forall m,n \in \mathbb{Z}$, and $\langle\ ,\ \rangle$ is a GT form iff K is a GTK, i.e.: $K(m + 1, n + 1) = K(m,n)$ only where $\text{sign}(m + 1) = \text{sign } m$ and $\text{sign}(n + 1) = \text{sign } n$.

Observe that, while, for a given form, the measure $\mu \geq 0$ in the Herglotz-Bochner theorem is unique, the matrix measure $\mu = [\mu_{ij}]$ in the GBT is not unique. In fact, μ_{11} and μ_{22} are uniquely determined but not so μ_{12}, and there are parametrization formulas, as well as Schur type constructions, describing all such $[\mu_{ij}]$ (cf. [Ar1,2], [Aℓ1,2]), that generalize the classical parametrization of the Nehari theorem given by Adamjan, Arov and Krein [AAK1].

In the special case when the weakly positive matrix measure $\nu = [\nu_{ij}]$ in Corollary 1(b) satisfies $\nu_{11} = \nu_{22} = (M + 1)\rho$, $\nu_{12} = \nu_{21} = (M - 1)\rho$, for a measure $\rho \geq 0$ in $\mathbf{T}$ and a fixed construct $M \geq 1$, the result leads to precised versions and new variants of the Helson-Szegö theorem for one or two weights, a related theorem of Koosis, and the prediction results of Helson-Sarason and Ibragimov-Rozanov (cf. [ArCS]).

The GBT provides also refinements to the Fefferman-Stein duality theory and to the study of Carleson measures (cf. [ArCS]), as well as to harmonizable stochastic processes (cf. [CS6]).

The results mentioned here are also proved valid for GT forms in $\mathbb{R}$, either by reduction to the case in $\mathbf{T}$ (as done in [ArC1] and [CS 9]), or through the method of local semigroups of isometries [Br], a notion appearing naturally in the consideration of GT forms, providing a unified treatment of the GT forms and the M.G. Krein extension method (cf. also [L]).

The lifting property of Corollary 1(b) has its own development in two directions. The first is suggested by the Adamjan-Arov-Krein (A-A-K) theorem [AAK2], and concerns the so-called conditionally positive scalar products, which are positive in certain subspaces of finite codimension. This generalization of the lifting theorem through conditionally positive extensions leads to weighted versions of the A-A-K theorem, as well as of the Helson-Szegö theorem [CS 17].

On the other hand, the restriction of a positive GT form to $P_1 \times P_2$ gives a Hankel form B_0: $P_1 \times P_2 \to \mathbb{C}$, while its restrictions to $P_1 \times P_1$ and $P_2 \times P_2$ give two Toeplitz quadratic seminorms, $\|\cdot\|_1$ and $\|\cdot\|_2$, such that the positivity of the form is equivalent to $|B_0(f,g)| \leq \|f\|_1 \|g_2\|$ for $f \in P_1$, $g \in P_2$. Corollary 1(a) expresses essentially that B_0 can be extended to a Toeplitz form $B: P \times P \to \mathbb{C}$, such that $|B(f,g)| \leq \|f\|_1 \|g\|_2$ for all $f,g \in P$, and $B = B_0$ in $P_1 \times P_2$. In this form, Corollary 1(a) is valid in the more general situation when P is replaced by an arbitrary pre-Hilbert space H, P_1 and P_2 by two subspaces W_1 and W_2, and the shift by a linear isomorphism σ of H onto H, such that $\sigma W_1 \subset W_1$ and $\sigma^{-1} W_2 \subset W_2$, and is a result on extension of σ-Hankel and essentially σ-Hankel forms (see Section 3).

Such abstract lifting theorems provide as immediate corollaries the intertwining lifting theorem of Sz.Nagy and Foias [Sz.N-F] and the interpolation theorem of Sarason [Sa], and new n-dimensional versions of such theorems (cf. [CS 16]).

Moreover, the GBT and Corollary 1 lead to new n-dimensional and noncommutative (quantized) versions of the theorems of Nehari, A-A-K and Helson-Szegö [CS 17,18], as well as to non-unitary versions of the Sz.Nagy and Foias theorem, closely related to the Grothendieck Inequality [CS 15].

There are also nonlinear lifting results [CS 13] connected with the local nonlinear commutant lifting theorem of Ball, Foias, Helson and Tannenbaum [BFHT] and to the multilinear analysis of Coifman and Meyer, as well as with ideas of Gelfand and Kostuchenko.

Finally, the weakly positive matrix measures have been applied to multivariate prediction problems [Do1,2], and to linear systems and colligations in Krein spaces [M].

2. THE GBT AND THE THEOREMS OF HELSON-SZEGÖ AND NEHARI.

In this section we consider in more detail the notion of positivity in P^2, and its application to the theorems of Helson-Szegö and Nehari. In particular, we show that the GBT (Theorem 1) can be expressed in terms of order properties that underlie the notion of complete positivity in C*-algebras (cfr. [Arv], [Da], [E]).

Given a $*$-algebra A, for each $n \in \mathbb{N}$, let $\mathbf{M}_n(A)$ be the set of the $n \times n$ matrices $f = [f_{ij}]$, with entries $f_{ij} \in A$, and let $\mathbf{M}_n(A')$ be the set of the $n \times n$ matrices $\ell = [\ell_{ij}]$, with entries $\ell_{ij}: A \to \mathbb{C}$, linear functionals in A. Then, $\mathbf{M}_n(A)$ is also a $*$-algebra under the algebraic matrix operations, and each $\ell \in \mathbf{M}_n(A')$ defines a linear functional in $\mathbf{M}_n(A)$ by

$$\ell(f) = \ell([f_{ij}]) = \sum_{i,j} \ell_{ij}(f_{ij}), \ \forall f = [f_{ij}] \in \mathbf{M}_n(A). \qquad (2.1)$$

An element $f \in A$ is said <u>positive in A</u> if $f = g^*g$, for some $g \in A$, and similarly for the $*$-algebra $\mathbf{M}_n(A)$. An $\ell \in \mathbf{M}_n(A')$ is said <u>positive</u> if $\ell(f) \geq 0$ whenever f is positive in $\mathbf{M}_n(A)$.

When $n = 2$, every $\ell \in \mathbf{M}_2(A')$ defines a sesquilinear form $\langle\langle\ ,\ \rangle\rangle$ in $A^2 = A \times A$, given by

$$\langle\langle (f_1,f_2),\ (g_1,g_2) \rangle\rangle := \ell([g_j^*f_i]) = \sum_{i,j} \ell_{ij}(g_j^*f_i) \qquad (2.2)$$

and the order properties of $\mathbf{M}_2(A')$ can be expressed in terms of scalar products.

Since P is an algebra of continuous functions with $*$-operation $f^*(t) = \overline{f(t)}$, taking $A = P$, the GBT relates with the order in $\mathbf{M}_2(P')$ as follows. The Fejer-Riesz lemma says that $f \in P$ satisfies $f(t) \geq 0$ for all t iff $f = g\bar{g}$, for some $g \in P_1$, and iff f is a positive element of P. In $\mathbf{M}_2(P)$ we have the following properties frequently used in the application of the GBT.

(A) For a given $f = [f_{ij}] \in \mathbf{M}_2(P)$ the following conditions are equivalent:

(A_1) f is a finite sum of positive elements of $\mathbf{M}_2(P)$.

(A_2) f is a finite sum of elements of the form $[f_i\bar{f}_j]$, $f_1, f_2 \in P$.

(A_3) $[f_{ij}(t)]$ is a positive matrix for all $t \in \mathbf{T}$.

(A_4) $\Sigma_{ij} f_{ij}\varphi_i\bar{\varphi}_j$ is a positive element of P for all pairs $\varphi_1\varphi_2 \in P$.

As a consequence follows

(B) $\ell = [\ell_{ij}] \in \mathbf{M}_2(P')$ is positive iff

$$\ell_{11}(f_1\bar{f}_1) \geq 0, \ \ell_{22}(f_2\bar{f}_2) \geq 0 \quad \text{and}$$
$$|\ell_{12}(f_1\bar{f}_2)| \leq \ell_{11}(f_1\bar{f}_1)^{1/2} \ \ell_{11}(f_1\bar{f}_2)^{1/2} \tag{2.3}$$

for every pair $f_1, f_2 \in P$, and iff there exists $\mu = [\mu_{ij}] \in \mathbf{M}_2^+(\mathbf{T})$ such that $\ell(f) = \mu(f)$ for all $f \in \mathbf{M}_2(P)$. (Here the measures μ_{ij} are considered as linear functionals in P, i.e., μ is considered as an element in $\mathbf{M}_2(P')$.)

(C) $\ell = [\ell_{ij}] \in \mathbf{M}_2(P')$ is weakly positive, i.e.: $\Sigma_{ij} \ell_{ij}(f_i\bar{f}_j) \geq 0$ whenever $f_1 \in P_1$, $f_2 \in P_2$, iff (2.3) holds for $f_1 \in P_1$ and $f_2 \in P_2$, and iff there exists a weakly positive $\nu = [\nu_{ij}] \in \mathbf{M}_2(\mathbf{T})$ such that $\ell(f) = \nu(f)$ whenever $f = [f_i\bar{f}_j]$ for $f_1 \in P_1$, $f_2 = P_2$.

(D) $\mu = [\mu_{ij}] \in \mathbf{M}_2(\mathbf{T})$, considered as an element of $\mathbf{M}_2(P')$ is positive in $\mathbf{M}_2(P')$ iff $\mu \in \mathbf{M}_2^+(\mathbf{T})$, and iff

$$\mu_{11} \geq 0, \ \mu_{22} \geq 0, \ \mu_{21} = \overline{\mu_{12}} \text{ and } |\mu_{12}(\Delta)|^2 \leq \mu_{11}(\Delta)\mu_{22}(\Delta) \tag{2.4}$$

for all Borel sets Δ in $\mathbf{T}$. In the special case when $\mu_{11} = \mu_{22} =$ the Lebesgue measure, μ is positive in $\mathbf{M}_2(P')$ iff $d\mu_{12} = \varphi(t)dt$, $d\mu_{21} = \overline{\varphi(t)}dt$, for $\varphi \in L^\infty$, $\|\varphi\|_\infty \leq 1$.

The last assertion is related to the following property in general C*-algebras: $a \in A$ satisfies $\|a\| \leq 1$ iff $[a_{ij}]$ is positive in $M_2(A)$ for $a_{11} = a_{22} = 1$, $a_{12} = a_{21}{}^* = a$ (cf. [E]).

Property (A) is a consequence of a corresponding property of general $*$-algebras (see [T], IV, 3), with the exception of the implication $(A_3) \Rightarrow (A_2)$ which follows from the more refined property:

PROPOSITION. ([CS 4], [CS 8]). *Let* $M_2^a(P) = \{f = [f_{ij}] \in M_2(P)\colon f_{12} \in P_1, f_{21} = \bar{f}_{12} \in P_2\}$. *If* $f = [f_{ij}] \in M_2^a(P)$ *is positive in* $M_2(P)$, *then* f *is the limit (in the norm of the supremum) of finite sums of elements of the form* $[f_i\bar{f}_j] \in M_2(P)$, *where* $f_1 \in P_1$, $f_2 \in P_2$. *Consequently, every weakly positive matrix measure* $\mu \in M_2(\mathbf{T})$, *considered as an element of* $M_2(P')$, *satisfies* $\mu(f) \geq 0$, *whenever* $f \in M_2^a(P)$ *and* f *is positive in* $M_2(P)$.

SKETCH OF PROOF. By $(A_1) \Rightarrow (A_3)$, $f_{11} \geq 0$, $|f_{12}|^2 \leq f_{11}f_{22}$, and, by Fejer-Riesz, $f_{11} = g_1\bar{g}_1$, $f_{22} = g_2\bar{g}_2$, for $g_1, g_2 \in P_1$. Since f_{12}, g_1 and g_2 are analytic, setting $g_i = \beta_i h_i$, where β_i is the Blashke product of g_i, this gives $f_{11} = h_1\bar{h}_1$, $f_{22} = h_2\bar{h}_2$ and $f_{12} = \gamma h_1 h_2$, where $\gamma_1 h_1$ and h_2 are continuous analytic functions and $|\gamma(t)| \leq 1$. Thus $f = [\varphi_i\bar{\varphi}_j] + [\psi_i\bar{\psi}_j]$, where $\varphi_1 = \gamma\varphi_1 = \varphi_2 = \bar{h}_2$, $\psi_1 = (1 - |\gamma|^2)h_1$, $\psi_2 = 0$. Since $\varphi_1, \varphi_2, \psi_1, \psi_2$ are limits of analytic polynomials, the assertion follows. □

The Proposition and property (A_3) are analogues of the Fejer-Riesz property in $M_2(P)$.

On the other hand, from the definition of Toeplitz and GT forms it is easy to deduce the two following properties, that can be adopted as definitions.

(E) Let $\langle\ ,\ \rangle$ be a Toeplitz form in P, and set $\ell(e_n) = \langle \ell_n, \ell_0 \rangle$, $\ell_n(t) = \exp(int)$. Then ℓ extends to a linear functional ℓ in P, such that $\langle f,g \rangle = \ell(f\bar{g})$, for $f,g \in P$.

(F) Let $\langle\ ,\ \rangle$ be a GT form in P, and set $\ell_{11}(e_{n-m}) = \langle \ell_n, \ell_m \rangle$ if $n \geq 0$, $m \geq 0$, $\ell_{22}(e_{n-m}) = \langle e_n, e_m \rangle$ if $n < 0$, $m < 0$, $\ell_{12}(e_{n-m}) = \langle e_n, e_m \rangle$ if $n \geq 0$, $m < 0$, $\ell_{21}(e_{n-m}) = \langle e_n, e_m \rangle$ if $n < 0$, $m \geq 0$. Then each ℓ_{ij} extends to a linear functional ℓ_{ij} in P, such that $\langle f,g \rangle = \ell_{ij}(f\bar{g})$ if $f \in P_i$, $g \in P_j$, for $i,j = 1,2$. The extensions ℓ_{11} and ℓ_{22} are unique, but not so ℓ_{12} and ℓ_{21}.

(G) From (F) and (C) follows that <u>the GBT can be deduced from either one of the lifting properties (a) and (b) of Corollary 1</u>. In fact, by (F), every positive GT form is given by a weakly positive matrix $\ell = [\ell_{ij}] \in M_2(P')$, which, by (C) can be replaced by a weakly positive matrix measure $\nu \in M_2(\mathbf{T})$, and, by the lifting property, ν can be replaced by a positive matrix measure $\mu \in M_2^+(\mathbf{T})$.

The Herglotz-Bochner theorem is known to be a consequence of the Fejer-Riesz lemma, property (E), and the Krein version of the Hahn-Banach theorem for positive linear functionals. Indeed, if $\langle\ ,\ \rangle$ is a positive Toeplitz form in P, then, by (E), $\langle f,g \rangle = \ell(f\bar{g})$, with $\ell(f\bar{f}) \geq 0$, so that, by Fejer-Riesz, ℓ is positive in P and extends to a positive linear functional in $C(\mathbf{T})$, furnishing the desired positive measure μ.

Similarly, <u>the GBT follows from properties (F), (C), (D) and the Krein-Hahn-Banach extension theorem</u>. Indeed, as observed above, it is enough to prove the lifting property (b) in Corollary 1. But, if μ is a weakly positive matrix measure, then, by (C) and (F), its restriction to the subspace $M_2^a(P)$ of $M_2(P)$ is positive in the subspace, extends to a positive functional μ in $M_2(P)$ by K-H-B, and, by (D), $\mu \in M_2^+(\mathbf{T})$.

<u>REMARK 3</u>. Here we consider scalar products in P and P^2 because of the simplicity of the functions $f \in P$. But since P is not a Banach algebra, to have a closer relation with Banach algebras, one should work, as Gohberg and his collaborators do, in the Wiener spaces W, W_1 and W_2, instead than in P, P_1 and P_2 (cf. [DG 12], [EG]).

An important special case of Corollary 1 is when the weakly positive matrix measure $\nu = [\nu_{ij}]$ is of the form $\nu_{11} = \nu_{22} = (M - 1)\rho$, $\nu_{12} = \nu_{21} = (M + 1)\rho$, for a positive measure ρ and a constant $M \geq 1$. In this case, condition (1.11) becomes:

$$\rho \text{ is absolutely continuous, } d\rho = \omega(t)dt \text{, and, for } h \in H^1,$$

$$2\omega(t) \leq \operatorname{Re}h(t) \leq |h(t)| \leq 2M^2\omega(t), \ |\arg h(t)| \leq \frac{\pi}{2} - \epsilon_M \text{ a.e.} \tag{2.5}$$

so that $\omega \sim |h|$ [CS 3].

If Γ is the Hilbert transform, i.e. the bounded operator in $L^2(\mathbf{T})$ satisfying $\Gamma f = -i(P_1 f - P_2 f)$ for all $f \in P$, then two bounded functions can be defined, $u = \log(\omega/|h|)$ and $v = \arg h$, with $\Gamma v = \log|h|$, since $\log h = \log|h| + i \arg h$ for $h \in H^1$. Then (2.5) is easily seen to be equivalent to each of the following:

$$\exists w \text{ real valued, such that } w \sim \omega \text{ and } |\Gamma w| \leq M^2 w \text{ a.e.} \tag{2.6}$$

$$\exists u,v \in L^\infty, \ \|u\|_\infty \leq C_M, \ \|v\|_\infty \leq \frac{\pi}{2} - \epsilon_M, \text{ such that } \omega = \exp(u + \Gamma v). \tag{2.7}$$

A classical problem in analysis was to characterize the measures $\rho \geq 0$ in $\mathbf{T}$ such that Γ be bounded in $L^2(\rho)$, with norm $\leq M$,

$$\int |f|^2 d\rho \leq M^2 \int |f|^2 d\rho, \ \forall f \in P. \tag{2.8}$$

Condition (2.8) can be rewritten as

$$\mu_{11}((P_1 f)(\overline{P_1 f})) + \mu_{12}((P_1 f)(\overline{P_2 f})) + \mu_{21}((P_2 f)(\overline{P_1 f}))$$

$$+ \mu_{22}((P_2 f)(\overline{P_2 f})) \geq 0 \tag{2.8a}$$

for

$$\mu_{11} = \mu_{22} = (M - 1)\rho, \ \mu_{12} = \mu_{21} = (M + 1)\rho. \tag{2.8b}$$

Thus the $\mu = [\mu_{ij}]$, for these μ_{ij}, is weakly positive and of the special type described above, so that the measures ρ are absolutely continuous, $d\rho = \omega dt$, and characterized by each of the conditions (2.5), (2.6) or (2.7). Condition (2.7) is the necessary and sufficient condition in the Helson-Szegö theorem, with further control of the constants, and (2.5) and (2.6) give other equivalent characterization of these measures ([CS 3,5]).

An important advantage of the argument above is that it solved the previously open problem of characterizing the pairs of positives measures ρ_1, ρ_2 such that $\Gamma: L^2(\rho_1) \to L^2(\rho_2)$, continuously, with control of the norm ([CS 4,5]).

Moreover, the characterizations (2.5), (2.6) extend to L^p, $1 < p < \infty$, as follows. (For more details on this subject, see [CS 14]).

COROLLARY 2. [CS 5]. *Γ acts continuously in $L^p(\rho)$, $1 < p < \infty$, with norm $\leq M$, iff $d\rho = \omega(t)dt$ and for all $0 \leq \varphi \in L^{p^*}$, $\|\varphi\| \leq 1$, there are $a \in L^{p^*}$, $\|a\| \leq C_M$, and $h \in H^1$, such that*

$$2\varphi\omega^{2/p} \leq \mathrm{Re}\, h \leq |h| \leq a w^{2/p} \text{ a.e.}, \tag{2.5a}$$

where $1/p^* = |1 - 2/p|$.

For $d\rho = dt$, $\omega \equiv 1$, Corollary 2 reduces to the classical M. Riesz Inequality for the Hilbert transform.

Let us see now that the GBT provides a refinement of the Nehari theorem on Hankel forms.

A Hankel form is a sesquilinear form $B: P_1 \to P_2$ satisfying any of the following equivalent conditions:

(i) $B(\sigma f, g) = B(f, \sigma^{-1}g)$, $\forall f \in P_1$, $g \in P_2$;

(ii) $\exists$ a linear functional $\ell: P_2 \to \mathbb{C}$, such that $B(f,g) = \ell(f\bar{g})$;

(iii) $\exists$ a sequence β: $\mathbb{Z}_+ \to \mathbb{C}$, such that $B(e_m, e_n) = \beta_{m-n}$ for all $m \geq 0$, $n < 0$, where $e_m(t) = \exp(int)$.

Observe that (i) makes sense because

$$\sigma P_1 \subset P_1 \quad \text{and} \quad \sigma^{-1} P_2 \subset P_2. \tag{2.9}$$

Every positive Toeplitz form B: $P \times P \to \mathbb{C}$ defines a Toeplitz quadratic seminorm by $\|f\|_B = B(f,f)^{1/2}$, which converts P into a pre-Hilbert seminormed space. If B_0: $P_1 \times P_2 \to \mathbb{C}$ is a Hankel form and B_1, B_2 are two positive Toeplitz forms, we write

$$B_0 \leq (B_1, B_2) \text{ in } P_1 \times P_2 \quad \text{iff } |B_0(f,g)| \leq \|f\|_{B_1} \|g\|_{B_2},$$

$$\forall f \in P_1,\ g \in P_2. \tag{2.10}$$

If $\langle\ ,\ \rangle$ is a positive GT form in P then, by (G), its restrictions to $P_1 \times P_1$ and $P_2 \times P_2$ coincide with positive Toeplitz forms B_1, B_2, while its restriction to $P_1 \times P_2$ is a Hankel form B_0, which, by Schwarz inequality, satisfies $B_0 \leq (B_1, B_2)$. Thus, it is the same to give a positive GT form in P or to give a Hankel form B_0 and two positive Toeplitz forms B_1 and B_2, satisfying $B_0 \leq (B_1, B_2)$ in $P_1 \times P_2$. From the GBT follows

COROLLARY 3.

(1) *A Hankel form* B_0: $P_1 \times P_2 \to \mathbb{C}$ *satisfies* $B_0 \leq (B_1, B_2)$ *in* $P_1 \times P_2$, *for* $B_1 \geq 0$, $B_2 \geq 0$, *Toeplitz forms in* P, *iff there exists a Toeplitz form* B: $P \times P \to \mathbb{C}$ *such that* $B = B_0$ in $P_1 \times P_2$ *and* $B \leq (B_1, B_2)$ in $P \times P$.

Moreover, setting $B(f,g) = \mu_{12}(f\bar{g})$, $\mu_{21} = \overline{\mu_{12}}$, $B_1(f,g) = \mu_{11}(f\bar{g})$, $B_2(f,g) = \mu_{22}(f\bar{g})$, *for all* $f,g \in P$, it is $[\mu_{ij}] \in \mathbf{M}_2^+(\mathbf{T})$.

(2) *In the special case when* $B_1(f,g) = B_2(f,g)$ *for all* $f,g \in P$,

$B(f,g) = \int f\bar{g}\varphi dt$, for $\varphi \in L^\infty$, $\|\varphi\|_\infty \leq 1$.

Part (2) of Corollary 3 is the classical Nehari theorem, and φ is called a symbol of B_0.

The notion of positive GT forms can be formulated in an abstract setting as follows. A pair of systems $[H_1;\sigma_1;W_1]$, $[H_2;\sigma_2;W_2]$ is an algebraic scattering structure if H_1 and H_2 are vector spaces, $\sigma_1\colon H_1 \to H_1$, $\sigma_2\colon H_2 \to H_2$ are linear isomorphisms, and $W_1 \subset H_1$, $W_2 \subset H_2$ are subspaces satisfying

$$\sigma_1 W_1 \subset W_1 \quad \text{and} \quad \sigma_2^{-1} W_2 \subset W_2. \tag{2.11}$$

For $H_1 = H_2$, a Hilbert space, and $\sigma_1 = \sigma_2$, a unitary operator in that space, the structure is a classical Lax-Phillips scattering system.
Let B_1 and B_2 be two fixed Toeplitz positive sesquilinear forms in H_1, H_2, respectively, i.e., forms satisfying

$$B_i(f,f) \geq 0,\ B_i(\sigma_i f,\ \sigma_i g) = B_i(f,g)\ \forall f,g \in H_i,\ i = 1,2. \tag{2.12}$$

An abstract Hankel form in an algebraic scattering structure is a sesquilinear form $B\colon W_1 \times W_2 \to \mathbb{C}$ satisfying $B(\sigma_1 f,g) = B(f,\sigma_2^{-1} g)$, $\forall f \in W_1$, $g \in W_2$, and an abstract GT form in such a structure is a Hankel form B satisfying $B \leq (B_1,B_2)$ in $W_1 \times W_2$, for B_1,B_2 as in (2.12).

In what follows, for simplicity it will be assumed that H_1 and H_2 are Hilbert spaces, B_1 and B_2, their scalar products, σ_1 and σ_2, unitary operators in H_1,H_2, respectively, and W_1 and W_2, closed subspaces satisfying

(2.11). In the next section it is shown that Corollary 3 extends to this abstract setting, but since the notions of this section relate to order properties of $*$-algebras that do not translate to the abstract situation, another type of ideas will be required.

3. GNS CONSTRUCTION, WOLD DECOMPOSITION AND ABSTRACT LIFTING THEOREMS.

The Herglotz-Bochner theorem is known to be closely related to the Gelfand-Naimark-Segal (GNS) principle. If $\langle\ ,\ \rangle$ is a positive Toeplitz form in P, then it converts P into a Hilbert space, i.e., there exist a Hilbert space H and a linear map J: $P \to H$ such that $\langle f,g\rangle = \langle Jf,Jg\rangle_H$, and JP is dense in H.

Setting $U(Jf) = J(\sigma f)$, U extends to a unitary operator in H, so that $\{n \mapsto U^n\}$ is a unitary representation of $\mathbb{Z}$, and $\xi = Je_0$ is a cyclic element, i.e., the elements $U^n\xi$ span H, and $\langle e_n,e_m\rangle = \langle U^n\xi,U^m\xi\rangle$. This is called the GNS construction, and setting $\mu(\Delta) := \langle E(\Delta)\xi,\xi\rangle$, $E(\Delta)$ the spectral measure of U, the Bochner theorem follows: $\langle f,g\rangle = \mu(f\bar{g})$.

If $\langle\ ,\ \rangle$ is a positive GT form, instead of a positive Toeplitz one, then it still defines a Hilbert space H′ and a map J: $P \to H'$, but now U′, given again by $U'(Jf) = J(\sigma f)$, need not be unitary. U′ is an isometry with domain $JP_1 + J(\sigma^{-1}P_2)$, and range $J(\sigma P_1) + J(P_2)$. This isometry extends to a unitary operator U in a larger space H.

THEOREM 2. <u>(GNS construction for GT forms</u> [ArCS]). *If $\langle\ ,\ \rangle$ is a positive GT form in P, then there exist a Hilbert space* H, *a unitary operator* U *in* H, *a linear map* J: $P \to H$ *and a cyclic pair* $\xi_1 = Je_0$, $\xi_2 = Je_{-1}$, $e_n(t) = \exp(int)$, *such that, for* $i,j = 1,2$,

$$\langle e_n,e_m\rangle = \langle U^n\xi_1,\ U^m\xi_2\rangle \text{ if } e_n \in P_i,\ e_m \in P_j. \tag{3.1}$$

Setting $\mu_{ij}(\Delta) = \langle E(\Delta)\xi_i,\xi_j\rangle$, *for* E *the spectral measure of* U, *the matrix measure* $\mu = [\mu_{ij}] \in M_2^+(T)$ *satisfies the GBT.*

REMARK 4. This theorem extends for operator-valued forms $\langle f,g\rangle \in L(N)$, N a Hilbert space, by replacing P by $P(N)$, the set of trigonometric polynomials $f(t) = \Sigma_n \xi_n(t)$, with coefficients $\xi_n \in N$, and defining in $P(N)$ the scalar product

$$\langle f,g\rangle := \sum_{n,m} \langle\langle e_n,e_m\rangle\xi_n,\xi_m\rangle_N. \tag{3.2}$$

The pair Je_0, Je_{-1} in Theorem 2 is cyclic because $P_1 \ominus \sigma P_1 = \{ce_0\}$, and $P_2 \ominus \sigma^{-1}P_2 = \{ce_{-1}\}$ are one-dimensional subspaces. Replacing P_1 and P_2 by the abstract W_1 and W_2 of an algebraic scattering structure (see the end of Section 2), then $R_1 = W_1 - \sigma W_1$, $R_2 = W_2 - \sigma^{-1}W_2$ need not be one-dimensional, but it is still a cyclic set.

In fact, the Wold-Kolmogorov decomposition gives, for i = 1,2,

$$H_i = \bigoplus_{n=-\infty}^{\infty} \sigma_i^n R_i \oplus H_i^0 \oplus H_i^1 \tag{3.3}$$

where

$$H_i^1 = \bigcap_{n\geq 0} \sigma_i^n W_i, \tag{3.3a}$$

so that $\sigma_i H_i^1 = H_i^1$ and $\sigma_i H_i^0 = H_i^0$, and every $f_i \in H_i$ can be written as

$$f_i = \bigoplus_{n=-\infty}^{\infty} \sigma_i^n f_{n,i} + f_i^0 + f_i^1, \quad f_{n,i} \in R_i, \; f_i^0 \in H_i^0, \; f_i^1 \in H_i^1, \tag{3.3b}$$

(cf. [CS]).

THEOREM 3. (Analogue of the GNS construction for the Hankel forms)

If $B: W_1 \times W_2 \to \mathbb{C}$ *is a Hankel form in an algebraic scattering structure* $[H_1;\sigma_1;W_1]$, $[H_2;\sigma_2;W_2]$ *with fixed positive Toeplitz forms* B_1 *and* B_2, *and* $B \leq (B_1,B_2)$ in $W_1 \times W_2$, *then there exist a unitary operator* U *in a Hilbert space* N, *and two linear maps,* $J_1: H_1 \to N$ *and* $J_2: H_2 \to N$, *such that,*

REMARK 5. As observed in Remark 2, it is sometimes more convenient to state these results in terms of kernels $K: \mathbb{Z} \times \mathbb{Z} \to \mathbb{C}$ instead of forms in P. In particular, for positive definite kernels K there is a GNS theorem as Theorem 2, and, as observed in Remark 4, this holds also for operator-valued kernels $K(m,n) \in L(N)$. If A is a C*-subalgebra of $L(H)$, H a Hilbert space, and K is a kernel assigning to each pair $m,n \in \mathbb{Z}$ a bounded linear operator $K(m,n)\colon A \to L(H)$, then this kernel is called completely positive definite if for every pair of finite sets $\{m_j\} \subset \mathbb{Z}$, $\{\xi_j\} \subset H$, and linear operators $\varphi_j\colon A \to L(H)$,

$$\Sigma_{jk} < \xi_j, K(n_j,n_k)(\varphi_j^* \varphi_k)\xi_k> \geq 0.$$

The GNS extends also to completely positive definite Toeplitz kernels, leading to an important theory. The corresponding development for GT kernels should be of independent interest.

REMARK 6. If K is a positive definite kernel in $\mathbb{Z} \times \mathbb{Z}$, so is exp K, and the second quantization in Fock spaces exp H is known to allow to obtain the GNS construction of exp K from that of K, and similarly for the generalized GNS construction of Theorem 2. Of special interest are the characterizations of the so-called S-positive definite Toeplitz kernels, corresponding to representations in exp H of this type, and having as cyclic element the vacuum exp 0. The analogue problem for GT kernels seems to be difficult and has still to be solved.

REMARK 7. For $H_1 = H_2$ and $\sigma_1 = \sigma_2$ = the identity, Theorem 3 reduces essentially to the Beatrous-Burbea [BeBu] generalization of a theorem of Bergman-Schiffer. The 2-parameter version of Theorem 3 given in the next section provides also a 2-parameter analogue of the Beatrous-Burbea result.

4. MULTIPARAMETER AND n-CONDITIONAL LIFTING THEOREMS, THE A-A-K THEOREM AND APPLICATIONS IN SEVERAL VARIABLES.

The lifting theorems of the previous two sections extend to multiparameter scattering structures. In order to avoid notational complications we develop here only the two parameter case, that reveals the

if f_1 *and* f_2 *are given by* (3.3b), *then* $J_1f_1 = \bigoplus_n U^n f_{n,1} + f_1^1$, $J_2f_2 = \bigoplus_n U^n f_{n,2} + f_2^1$, *and such that, if* $B'(f_1,f_2) := \langle J_1f_1, J_2f_2\rangle_N$, *for all* $f_1 \in H_1$, $f_2 \in H_2$, *then* $B'\colon H_1 \times H_2 \to \mathbb{C}$ *is a sesquilinear form satisfying* $B' = B$ *in* $V_1 \times V_2$, *and* $B' \leq (B_1,B_2)$ *in* $H_1 \times H_2$.

Moreover, if E *is the spectral measure of* U, *and, for each pair* $f_1 \in V_1$, $f_2 \in V_2$, $\mu_{f_1,f_2}(\Delta) := \langle E(\Delta)J_1f_1, J_2f_2\rangle_N$, *then* B' *admits the integral representation*

$$B(f_1,f_2) = \sum_{n,m} \int e_{n-m}(t)d\mu_{f_{n,1},f_{m,2}} + \sum_m \int e_{-m}(t)d\mu_{f_1^1,f_{m,2}} + \sum_n \int e_n(t)d\mu_{f_2^1,f_{n,1}} + \int e_0(t)d\mu_{f_1^1,f_2^1}. \tag{3.4}$$

This integral representation can be precised further, but we shall not go into the matter here.

When H_1 and H_2 are Hilbert spaces under the scalar products B_1 and B_2, then Theorem 3 can be extended further, replacing the condition (2.11) by the weaker condition

$$H_i = V_i \oplus W_i^+ \oplus W_i^-, \text{ with } \sigma_i W_i^+ \subset W_i^+,\ \sigma_i^{-1} W_i^- \subset W_i^- \tag{3.5}$$

i = 1,2, and the Hankel forms by the so called <u>essentially Hankel</u> forms B: $V_1 \times V_2 \to \mathbb{C}$, that satisfy

$$B(P_1\sigma_1 f_1,f_2) = B(f_1,P_2\sigma_2^{-1}f_2),\ \forall f_1 \in V_1,\ f_2 \in V_2, \tag{3.6}$$

where P_i is the orthoprojector of H_i onto V_i, i = 1,2.

The <u>general abstract lifting theorem</u> thus obtained for essentially Hankel forms is logically equivalent to the Sz.Nagy-Foias lifting theorem for intertwining contractions, and contains the Sarason theorem as a special case [CS 16]. Moreover, as stated in Section 4, this lifting theorem for Hankel forms extends to 2-parameter scattering and other abstract settings.

essential features of the extension, so that the multiparameter statements will be apparent.

For simplicity, we consider in this section the <u>scattering structures</u> $[H_i; \sigma_i; W_i]$, i = 1,2, where H_1,H_2 are Hilbert spaces under scalar products B_1,B_2, respectively, $\|f\|_i = B_i(f,f)^{1/2}$, i = 1,2, and $\sigma_1 \in L(H_1)$, $\sigma_2 \in L(H_2)$ are unitary operators satisfying (2.11).

We consider here the extension of the Lifting Theorem 3 to the case when there is another pair $\tau_1 \in L(H_1)$, $\tau_2 \in L(H_2)$ acting in the scattering structure, satisfying

$$\tau_1 W_1 \subset W_1 \quad \text{and} \quad \tau_2^{-1} W_2 \subset W_2, \tag{4.1}$$

and the commutation condition

$$\tau_i \sigma_i = e^{i\alpha} \sigma_i \tau_i \quad \text{for some } \alpha \in \mathbb{R},\ i = 1,2. \tag{4.2}$$

The subspaces

$$W_i^\sigma = \{f \in W_i : \sigma_i^k f \in W_i,\ \forall k \in \mathbb{Z}\}$$

and

$$W_i^\tau = \{f \in W_i : \tau_i^k f \in W_i,\ \forall k \in \mathbb{Z}\}$$

for i = 1,2, play an important role and, in what follows, it is assumed that

$$W_2 = \{f + g : f \in W_2^\sigma,\ g \in W_2^\tau\} \tag{4.4}$$

A sesquilinear form $B: H_1 \times H_2 \to \mathbb{C}$ is called <u>Toeplitz in the scattering structure</u> $[H_i; \sigma_i, \tau_i, W_i]$, i = 1,2, if, for all $f_1 \in H_1$, $f_2 \in H_2$,

$$B(\sigma_1 f_1, \sigma_2 f_2) = B(f_1, f_2) \quad \text{and} \quad B(\tau_1 f_1, \tau_2 f_2) = B(f_1, f_2). \tag{4.5}$$

Similarly, a form $B: W_1 \times W_2 \to \mathbb{C}$ is called <u>Hankel in the scattering structure</u> if, for all $f_1 \in W_1$, $f_2 \in W_2$,

$$B(\sigma, f_1, f_2) = B(f_1, \sigma_2^{-1} f_2) \text{ and } B(\tau_1 f_1, f_2) = B(f_1, \tau_2^{-1} f_2). \qquad (4.5a)$$

In this context, $B \le (B_1, B_2)$ means that $|B(f_1,f_2)| \le \|f_1\|_1 \, \|f_2\|_2$, i.e., that B is bounded with norm $\|B\| \le 1$.

Given a bounded form $B: W_1 \times W_2 \to \mathbb{C}$ we call a pair of forms $B': H_1 \times H_2 \to \mathbb{C}$, $B'': H_1 \times H_2 \to \mathbb{C}$ a <u>lifting pair for B</u> if $\|B'\| \le \|B\|$, $\|B''\| \le \|B\|$ and

$$B' = B \text{ in } W_1 \times W_2^{\sigma} \text{ and } B'' = B \text{ in } W \times W_2^{\tau}. \qquad (4.6)$$

THEOREM 4. (<u>Two-parameter lifting theorem in scattering structure</u> [CS 10,12], cf. also [CS 13]). *Given a scattering structure* $[H_1;\ \sigma_1, \tau_1;\ W_1]$, $[H_2;\ \sigma_2, \tau_2;\ W_2]$, *as described above, every bounded Hankel form* B *has a lifting pair of Toeplitz forms* B', B''.

This theorem applies, in particular, to the structure given in $L^2(\mathbf{T}^d)$, $d \ge 2$, by the shift operators, in the three following special cases, providing versions of the Nehari and Helson-Szegö theorems in several dimensions.

<u>Special case I</u>. $H_1 = H_2 = L^2(\mathbf{T}^2)$, $\sigma_1 = \sigma_2 = \sigma$, defined by $\sigma f(x,y) = e^{ix} f(x,y)$, $\tau_1 = \tau_2 = \tau$, defined by $\tau f(x,y) = e^{iy} f(x,y)$, $W_1 = H^2(\mathbf{T}^2) = \{f \in L^2: \hat{f}(m,n) = 0 \text{ if } m < 0 \text{ or } n < 0\}$, $W_2 = W_1^{\perp}$. Here $W_2^{\sigma} = \{f \in L^2: \hat{f}(m,n) = 0 \text{ if } n \le 0\}$, $W_2^{\tau} = \{f \in L^2: \hat{f}(m,n) = 0 \text{ if } m \ge 0\}$.

<u>Special case II</u>. $H_1 = H_2 = L^2(\mathbf{T}^2)$, $\sigma_1 = \sigma_2 = \sigma$, $\tau_1 = \tau_2 = \tau$, as above, $W_1 = \{f \in L^2: \hat{f}(m,n) = 0 \text{ if } n < 0\}$, $W_2 = \{f \in L^2: \hat{f}(m,n) = 0 \text{ if } n \ge 0\}$. Here $W_2^{\sigma} = W_2$ and $W_2^{\tau} = \{0\}$.

<u>Special case III</u>. $H_1 = H_2 = L^2(\mathbf{T}^2;\ \omega(x,y)dxdy)$, for some positive integrable weight ω. W_1 and W_2, subspaces of $L^2(\mathbf{T}^2;\omega)$, are defined as in I

or II.

The Hankel forms in special case I are called big Hankel forms in $L^2(T^2)$, and those in special case II, middle Hankel forms. The Nehari theorem for big Hankel forms can be stated in terms of $BMO(T^2)$ functions, in the sense of A. Chang and R. Fefferman (cf. [CS 19]).

Every bounded Hankel form $B: W_1 \times W_2 \to \mathbb{C}$ is associated with a bounded Hankel operator $T: W_1 \to W_2$, through

$$B(f_1,f_2) = \langle Tf_1,f_2\rangle, \forall f_1,f_2\rangle \forall f_1 \in W_1, f_2 \in W_2. \tag{4.7}$$

For every $n \in \mathbb{N}$, the singular numbers of B are defined by

$$s_n(B) := s_n(T) = \inf\{\|T - T_n\|: \text{rank } T_n \leq n\}$$

$$= \inf\{\|T|_E\|: E \text{ subspace of } W_1, \text{ codim } E \leq n\} \tag{4.8}$$

In particular, $s_0(B) = s_0(T) = \|T\| = \|B\|$.

For a bounded form $B: W_1 \times W_2 \to \mathbb{C}$, acting in a scattering structure $[H_i; \sigma_i; W_i]$, $i = 1,2$, a form $B^{(n)}$ is called an n-conditional lifting of B, if there is a subspace $M \subset W_1$ of codimension $\leq n$, such that

$$B^{(n)} = B \text{ in } M \times W_2 \text{ and } \|B^{(n)}\| = s_n(B). \tag{4.9}$$

THEOREM 5. (n-conditional lifting theorem in one-parameter scattering structures [CS 17]).

Given a scattering structure $[H_1; \sigma_1; W_1]$, $[H_2; \sigma_2; W_2]$, *every bounded Hankel form* B *has, for each* $n \in \mathbb{N}$, *an n-conditional Toeplitz lifting* $B^{(n)}$. *Hence, for each* n,

$$s_n(B) = \inf\{\|T - T_n\|: T_n \text{ Hankel operator of rank } \leq n\} \tag{4.10}$$

where T *is the Hankel operator given by* (4.7).

REMARK 7. In the case $H_1 = H_2 = L^2(\mathbf{T})$, $\sigma_1 = \sigma_2 = \sigma$, the shift, $W_1 = H^2(\mathbf{T})$, $W_2 = H^2_-(\mathbf{T})$, Theorem 5 reduces to the classical Adamjan-Arov-Krein (A-A-K) theorem [AAK2]. Theorem 5 which is based on a result in [Tr], also applies in the case $H_1 = L^2(\mathbf{T};\mu_1)$, $H_2 = L^2(\mathbf{T};\mu_2)$, $W_1 = H^2(\mathbf{T};\mu_1)$, $W_2 = H^2_-(\mathbf{T};\mu_2)$, providing two-weighted versions of the A-A-K theorem [CS 17].

Combining Theorems 4 and 5 follows

THEOREM 6. (Estimates for the singular numbers of Hankel operators in the bidimensional torus [CS 17]). *Let* T: $W_1 \to W_2$ *be a bounded Hankel operator,* W_1 *and* W_2 *be as in special cases* I *or* II. *Then for every* $n \in \mathbb{N}$,

$$\|P_\sigma T\| = s_0(P_\sigma T) \equiv s_n(P_\sigma T) \leq s_n(T)$$

and (4.11)

$$\|P_\tau T\| = s_0(P_\tau T) \equiv s_n(P_\tau T) \leq s_n(T),$$

where P_σ *and* P_τ *are the orthogonal projections of* W_2 *onto* W_2^σ *and* W_2^τ, *respectively. In particular, every compact Hankel operator in* $L^2(\mathbf{T}^2)$ *is zero.*

The abstract multiparameter lifting theorem also provides the extension of the Helson-Szegö theorem (see Section 2) to the multidimensional torus. In fact, in the circle, the theorem asserts that the Hilbert transform Γ is bounded in $L^2(\mathbf{T};\omega)$ iff $\log \omega \in BMO(\mathbf{T})$, with norm comparable to the operator norm of Γ. In [CS 12,14] it is proved that the double Hilbert transform $\Gamma = \Gamma_x\Gamma_y$ is bounded in $L^2(\mathbf{T}^2;\omega)$ iff $\log \in BMO(\mathbf{T}^2)$, with norm comparable to that of the operator norm of Γ. This result also extends, through the concept of simultaneously u-boundedness of operators, to the characterization of the weights ω for which Γ acts continuously in $L^p(\mathbf{T}^2;\omega)$ for $1 < p < \infty$ [CS 14].

Finally, the preceding results extend to $\mathbb{R}$ and $\mathbb{R}^d$, $d \geq 1$, as well as to the symplectic space $(\mathbb{C}^d, [\ ,\ ])$, where the shifts are replaced by the so-called twisted shifts in $\mathbb{R}^{2d}$. This provides noncommutative versions of Theorems 2, 5 and 6, and corresponding theorems in Fock spaces [CS 12,18].

REFERENCES

[AAK1] V.M. Adamjan, D.Z. Arov and M.G. Krein, Infinite Hankel matrices and generalized Caratheodory-Fejer and I. Schur problems, Func. Anal. Appl.., 2 (1968), 1-117 (in Russian).

[AAK] ____________________, Analytic properties of Schmidt pairs of a Hankel operator and generalized Schur-Takagi Problem, Mat. Sbornik 86 (1971), 33-73.

[Aℓ1] P. Alegria, On the Adamjan-Arov-Krein and Arocena parametrizations: a constructive and generalized version, Acta Cientif. Venez. 39 (1988), 107-116.

[Aℓ2] _________, Doctoral Dissertation, Univ. del Pais Vasco, 1992.

[Ar 1] R. Arocena, On generalized Toeplitz kernels and their relation with a paper of Adamjan, Arov and Krein, North Holland Math. Studies, 86 (1984), 1-22.

[Ar 2] _________, On a geometric interpretation of Schur parameters, Rev. U.M.A. 34 (1988), 150-165.

[Ar C] R. Arocena and M. Cotlar, Continuous Generalized Toeplitz kernels in $\mathbb{R}$, Portugaliae Math. 39 (1980) 419-434.

[Ar CS] R. Arocena, M. Cotlar and C. Sadosky, "Weighted inequalities in L^2 and lifting properties, Adv. Math. Suppl. Studies, 7A (1981), 95-128.

[Arv] W.B. Arveson, Subalgebras of C*-algebras, Acta Math. 123 (1969), 141-22.

[BFHT] J.A. Ball, C. Foias, J.W. Helton and A. Tannenbaum, On a local nonlinear commutant lifting theorem, Indiana Univ. Math. J. 36 (1987), 693-707.

[BeBu] F. Beatrous and J. Burbea, Positive-definiteness and its applications to interpolation problems of holomorphic functions, Trans. Amer. Math. Soc. 284 (1984), 247-270.

[Br] R. Bruzual, Local semigroups of contractions and some applications to Fourier representation theorems, Int. Eqs. & Op. Theory 10 (1987), 780-801.

[CS 1] M. Cotlar and C. Sadosky, A moment theory approach to the Riesz theorem on the conjugate function for general measures, Studia, Math., 53 (1975), 75-101.

[CS 2] ______________________ , Transformee de Hilbert theoreme de Bochner et le probleme des moments, I, C.R. Acad. Sci., Paris, A 285 (1977), 433-435, 661-665.

[CS 3] ______________________ , On the Helson-Szegö theorem and a related class of modified Toeplitz kernels, Proc. Symp. Pure Math. AMS, 34 (1979), 383-407.

[CS 4] ______________________ , Characterization of two measures satisfying the Riesz inequality for the Hilbert transforms in L^2, Acta Cient. Venez., 30 (1979), 346-348.

[CS 5] ______________________ , On some L^p versions of the Helson-Szegö theorem, in Conf. Harmonic Analysis in honor of A. Zygmund (Eds.: W. Beckner, A.P. Calderon, R. Fefferman, P.W. Jones), Wadsworth Int. Math. Series (1982), 306-317.

[CS 6] ______________________ , Generalized Toeplitz kernels, stationarity and harmonizability, J. Anal. Math., 44 (1985), 117-133.

[CS 7] ______________________ , A lifting theorem for subordinated invariant kernels, J. Func. Anal., 67 (1986), 345-359.

[CS 8] ______________________ , Lifting properties, Nehari theorem and Paley Iacunary inequality, Rev. Mat. Iberoamericana, 2 (1986), 55-71.

[CS 9] ______________________ , Prolongements des formes de Hankel generalisees en formes de Toeplitz, C.R. Acad. Sci., Paris, 1 305 (1987), 167-170.

[CS 10] ______________________, Nonlinear liftings theorems, integral representations and stationary processes in algebraic scattering systems, Operator Th.: Adv. & Appl. 41 (1989), 97-123.

[CS 11] ______________________, The Generalized Bochner theorem in algebraic scattering structures, in Analysis at Urbana, II, Cambridge University Press, London, 1989.

[CS 12] ______________________, Two-parameter lifting theorems and double Hilbert transforms in commutative and non-commutative settings, J. Math. Anal. & Appl. 150 (1990), 439-480.

[CS 13] ______________________ , Nonlinear lifting theorems, Integral representations and stationary processes in algebraic scattering systems, Op. Th.: Adv. & Appl. 41 (1989), 97-123.

[CS 14] ______________________, The Helson-Szegö Theorem in L^p of the bidimensional torus, Contemporary Math. 107 (1990), 19-37.

[CS 15] ______________________, Lifting of Hankel forms bounded by non Toeplitz norms, Int. Eqs. & Op. Theory, 14 (1991), 501-532.

[CS 16] ______________________, Transference of metrics induced by unitary couplings, J. Funct. Anal. to appear.

[CS 17] ______________________, Weighted and two-dimensional Adamjan-Arov-Krein theorems and Sarason commutators, Institut Mittag-Leffler Report No. 24, 1990/91.

[CS 18] ______________________, The Adamjan-Arov-Krein theorems in abstract and regular representations of $\mathbb{R}^2$ and the symplectic plane, Institut Mittag-Leffler Report No. 34, 1990/91.

[CS 19] ______________________, preprint.

[Da] E.B. Davies, Quantum Theory of Open Systems, Academic Press, London, New York, 1976.

[Do 1] M. Domiguez, A matricial extension of the Helson-Sarason Theorem and a characterization of some multivariate linearly complete regular processes, J. Multivariate Anal. 31 (1989) 289-310.

[Do 2] ___________, Multivariate prediction theory and weighted inequalities, Op. Th. Adv. & Appl. (1990).

[DG 1] H. Dym and I Gohberg, Extensions of matrix valued functions and block matrices, Indiana Univ. Math. J. 31 (1982), 733-765.

[DG 2] ____________________, Unitary interpolants, factorization indices and infinite Hankel block matrices, J. Funct. Anal. 54 (1983), 229-289.

[E] E.G. Effros, Aspects of non-commutative order, Lecture Notes in Math. 650, Springer-Verlag, Berlin, New York, 1970.

[EℓG] R. Ellis and I. Gohberg, Orthogonal systems related to infinite Hankel matrices, J. Funct. Analysis, to appear.

[HS] H. Helson and G. Szegö, A problem in prediction theory, Ann. Math. Pure Appl., 51 (1960), 107-138.

[JPR] S. Janson, J. Peetre and R. Rochberg, Hankel forms and Fock spaces, Rev. Mat. Ibero Amer. 8 (1987), 61-138.

[M] S. Marcantognini, Unitary colligations in Krein spaces, J. Operator Th. (1990).

[Mo 1] M.D. Moran, On Intertwining Dilations, J. Math. Anal. Appl. 141 (1989), 219-234.

[Mo 2] ________, On Commuting Isometries, J. Operator Th. 24 (1990), 75-83.

[Ne] Z. Nehari, On bilinear forms, Ann. of Math. 68 (1957), 153-162.

[Ni] N.K. Nikolskii, Treatise on the Shift Operator, Springer-Verlag, Berlin-Heidelberg-New York, 1986.

[Sa] D. Sarason, Generalized interpolation in H^∞, Trans. Amer. Math. Soc. 127 (1967), 179-203.

[Sz. N-F] B. Sz. Nagy and C. Foias, Analyse harmonique des operateurs de l'espace de Hilbert, Masson-Akad. Kiado, Paris and Budapest, 1970.

[T] N. Takesaki, Operator Algebras, Springer-Verlag, Berlin, New York, 1979.

[Tr] S. Treil, The theorem of Adamjan-Arov-Krein: Vector variant, Publ. Seminar LOMI Leningrad 141 (1985), 56-72. (In Russian).

M. Cotlar
Facultad de Ciencias
Universidad Central de Venezuela
Caracas 1040, Venezuela

C. Sadosky
Department of Mathematics
Howard University
Washington, D.C. 20059, USA

Operator Theory:
Advances and Applications, Vol. 58

REDUCTION OF THE ABSTRACT FOUR BLOCK PROBLEM TO A NEHARI PROBLEM

J.A. Ball, I. Gohberg and M.A. Kaashoek

The abstract version of the four block extension problem is reduced to an abstract Nehari problem of which the solution may be obtained via the band method. It is also shown that the maximum entropy solution of the four block problem may be obtained in this way from the maximum entropy solution of the associated Nehari problem.

0. INTRODUCTION

This paper concerns the abstract version of extension problems of the following type. Consider the 2×2 matrix function

$$\Phi(t) = \begin{pmatrix} f(t) & g(t) + \varphi(t) \\ h(t) & k(t) \end{pmatrix}, \quad t \in \mathbf{R}, \tag{0.1}$$

where f, g, h and k are given functions in $L_1(\mathbf{R})$. The problem is to find $\varphi \in L_1(\mathbf{R})$ such that $\varphi(t) = 0$ a.e. on $-\infty < t < 0$ and

$$\sup_{\lambda \in \mathbf{R}} \|\widehat{\Phi}(\lambda)\| < 1. \tag{0.2}$$

Here $\widehat{\Phi}$ denotes the Fourier transform of Φ, and the norm in (0.2) is the usual operator norm for matrices. The above problem, which is usually referred to as a four block problem, the analogous problem with the real line replaced by the circle, and other matrix-valued versions of the problem appeared in the middle of the eighties in mathematical system theory in connection with problems of H_∞-control (see [1], [2] and [5]). The existence of a solution (also for the optimal case) was proved in [4] by using the commutant lifting theorem. In a more general setting (involving many blocks) a description of all solutions has been given in [3] (see also [6]). In the paper [3] also the maximum entropy solution was singled out, and the special role played by the maximum entropy solution was clarified.

The four block problem has been treated in [10] by means of the band method, and the same approach was used in [13] to solve other versions of the four block problem. For rational matrix-valued functions an alternative solution based on an all pass embedding method may be found in [9]. Another direction in the theory of the four block problem concerns the computation of the minimal norm, i.e., the smallest possible value of the left hand side of (0.2) for all possible choices of φ in (0.1). The papers [7] and [8] develop a skew Toeplitz theory to treat such issues.

In this paper we solve an abstract version of the four block problem by reducing it to an abstract Nehari problem of which the solution may be obtained via the band method (see [10] - [13]). The method of solution described in the present paper may be applied to many concrete cases. For the problem (0.1), (0.2) discussed above it yields the following result. First, note that for the problem (0.1), (0.2) to be solvable it is necessary that

$$\text{(i)}\ \ \Delta(\lambda) := 1 - |\widehat{f}(\lambda)|^2 - |\widehat{h}(\lambda)|^2 > 0,$$
$$\text{(ii)}\ \ \delta(\lambda) := 1 - |\widehat{h}(\lambda)|^2 - |\widehat{k}(\lambda)|^2 > 0.$$

Assuming that (i) and (ii) hold, introduce the functions

$$\widehat{x}_1(\lambda) = \exp \widehat{y}_1(\lambda), \quad \widehat{x}_2(\lambda) = \exp \widehat{y}_2(\lambda),$$

where

$$y_1(t) = \begin{cases} \{\log(1 + \Delta(\lambda)^{-1}|\widehat{f}(\lambda)|^2)\}^\vee(t), & t > 0, \\ 0, & t < 0, \end{cases}$$

$$y_2(t) = \begin{cases} \{\log(1 + \delta(\lambda)^{-1}|\widehat{k}(\lambda)|^2)\}^\vee(t), & t > 0, \\ 0, & t < 0, \end{cases}$$

let

$$\widehat{w}(k) = \{1 + \Delta(\lambda)^{-1}|\widehat{f}(\lambda)|^2\}\Delta(\lambda)^{-1}\widehat{f}(\lambda)\overline{\widehat{h}(\lambda)}\widehat{k}(\lambda),$$

and consider

$$\gamma(t) = \{\widehat{x}_1(\lambda)(\widehat{g}(\lambda) + \widehat{\omega}(\lambda))\widehat{x}_2(\lambda)\}^\vee(t), \quad t \in \mathbf{R}.$$

Here $\ell^\vee$ denotes the inverse Fourier transform of the function ℓ. Then there exists Φ as in (0.1) such that (0.2) holds if and only if (i) and (ii) are satisfied and the Hankel operator

Γ on $L_2([0,\infty))$ defined by

$$(\Gamma u)(t) = \int_0^\infty \gamma(-t-s)u(s)\,ds, \quad 0 \le t < \infty,$$

has norm strictly less than one. In the latter case all solutions of the problem (0.1), (0.2) are obtained by taking

$$\varphi(t) = \{\widehat{x}_1(\lambda)^{-1}\widehat{z}(\lambda)\widehat{x}_2(\lambda)^{-1} - \widehat{g}(\lambda) - \widehat{\omega}(\lambda)\}^\wedge(t), \quad t > 0, \tag{0.3}$$

where z is any element of $L_1(\mathbf{R})$such that

$$z(t) = \gamma(t) \quad (t < 0), \qquad \sup_{\lambda\in\mathbf{R}} |\widehat{z}(\lambda)| < 1. \tag{0.4}$$

The proof of our main theorem uses elements of the band method (see [10] - [13]). The band method may also be applied directly to the four block problem considered here. However, for examples of interest, the present adaption of the band method leads to simpler formulas at the end. We also show that the triangular extension (see the next section for the definition) of the four block problem corresponds to the triangular extension of the associated Nehari problem. For the concrete example mentioned above the latter means that the maximum entropy solution for the problem (0.1), (0.2) is obtained by taking z in (0.3) to be the maximum entropy solution of the Nehari problem (0.4).

The present paper consists of two sections (not counting this introduction). The first section contains the main theorems; the proofs are given in the second section.

1. MAIN THEOREMS

To state the main theorems we need some notation and terminology. Throughout this paper $\mathcal{N}$ is an algebra with the following *triangular structure*:

$$\mathcal{N} = \mathcal{N}_\ell^0 \dot{+} \mathcal{N}_d \dot{+} \mathcal{N}_u^0. \tag{1.1}$$

The latter means that $\mathcal{N}$ is an algebra with an involution *, a unit e, and the spaces appearing in the decomposition (1.1) are subalgebras and have the following properties:

(i) $e \in \mathcal{N}_d = \mathcal{N}_d^*$,

(ii) $(\mathcal{N}_\ell^0)^* = \mathcal{N}_u^0$,

(iii) $\mathcal{N}_d\mathcal{N}_\alpha^0 = \mathcal{N}_\alpha^0\mathcal{N}_d = \mathcal{N}_\alpha^0 \quad (\alpha = \ell, u)$.

We put

$$\mathcal{N}_\ell = \mathcal{N}_\ell^0 \dot{+} \mathcal{N}_d, \quad \mathcal{N}_u = \mathcal{N}_d \dot{+} \mathcal{N}_u^0. \tag{1.2}$$

We write $P^0_{\mathcal{N}_\ell}$ (resp., $P_{\mathcal{N}_\ell}$) for the projection of $\mathcal{N}$ onto $\mathcal{N}_\ell^0$ (resp., $\mathcal{N}_\ell$) along the space $\mathcal{N}_u$ (resp., $\mathcal{N}_u^0$). In a similar way by interchanging the roles of ℓ and u, one defines the projections $P^0_{\mathcal{N}_u}$ and $P_{\mathcal{N}_u}$. By $P_{\mathcal{N}_d}$ we denote the projection of $\mathcal{N}$ onto $\mathcal{N}_d$ along the space $\mathcal{N}_\ell^0 \dot{+} \mathcal{N}_u^0$.

We shall also assume that $\mathcal{N}$ satisfies the following additional requirements:

(E0) The principal submatrices of a positive definite element in $\mathcal{N}^{4\times4}$ are positive definite in the corresponding algebras.

(E1) The positive definite elements in $\mathcal{N}$ admit a left and a right spectral factorization relative to the decomposition (1.1).

Recall that an element b in a *-algebra $\mathcal{A}$ is called *positive definite* (notation: $b > 0$) if $b = c^*c$ with c invertible in $\mathcal{A}$. Furthermore, $b \in \mathcal{N}$ is said to admit a *left (resp., right) spectral factorization* relative to (1.1) if b factors as $b = c^*c$ where c is an invertible element of $\mathcal{N}$ such that c and its inverse c^{-1} are both in $\mathcal{N}_\ell$ (resp., $\mathcal{N}_u$). Elements in $\mathcal{N}$ that admit a spectral factorization are automatically positive definite in $\mathcal{N}$. Axiom (E1) requires that the converse implication is also true.

We say that an element $\psi_\ell \in \mathcal{N}_\ell$ has a *strictly contractive extension* φ if

(a) $\varphi - \psi_\ell \in \mathcal{N}_u^0$,

(b) $e - \varphi^*\varphi > 0$.

The problem to find all (strictly) contractive extensions of φ_ℓ is called a *Nehari problem*.

Put $\mathcal{B} = \mathcal{N}^{2\times2}$, and set $E = \begin{pmatrix} e & 0 \\ 0 & e \end{pmatrix}$. We call an element $G \in \mathcal{B}$ *strictly contractive* if $E - G^*G$ is positive definite in $\mathcal{B}$.

THEOREM 1.1. *Given*

$$\Phi = \begin{pmatrix} \varphi_{11} & \varphi_\ell \\ \varphi_{21} & \varphi_{22} \end{pmatrix} \in \mathcal{B} = \mathcal{N}^{2\times2}, \quad \varphi_\ell \in \mathcal{N}_\ell, \tag{1.3}$$

there exists $x \in \mathcal{N}_u^0$ *such that*

$$G = \begin{pmatrix} \varphi_{11} & \varphi_\ell + x \\ \varphi_{21} & \varphi_{22} \end{pmatrix} \tag{1.4}$$

is strictly contractive if and only if

$$\Delta := e - \varphi_{11}^* \varphi_{11} - \varphi_{21}^* \varphi_{21} > 0, \tag{1.5}$$

$$\delta := e - \varphi_{21} \varphi_{21}^* - \varphi_{22} \varphi_{22}^* > 0, \tag{1.6}$$

and the element $\psi_\ell := P_{\mathcal{N}_\ell}\{x_1(\varphi_\ell + \omega)x_2\}$ *has a strictly contractive extension in* $\mathcal{N}$. *Here*

$$x_1^{\pm 1} \in \mathcal{N}_u, \quad e + \varphi_{11}\Delta^{-1}\varphi_{11}^* = x_1^* x_1, \tag{1.7a}$$

$$x_2^{\pm 1} \in \mathcal{N}_u, \quad e + \varphi_{22}^*\delta^{-1}\varphi_{22} = x_2 x_2^*, \tag{1.7b}$$

$$\omega = (e + \varphi_{11}\Delta^{-1}\varphi_{11}^*)^{-1}\varphi_{11}\Delta^{-1}\varphi_{21}^*\varphi_{22}. \tag{1.7c}$$

Furthermore, in that case all $x \in \mathcal{N}_u^0$ *such that* G *in* (1.4) *is strictly contractive are given by*

$$x = x_1^{-1}[y - P_{\mathcal{N}_u^0}\{x_1(\varphi_\ell + \omega)x_2\}]x_2^{-1}, \tag{1.8}$$

where $\psi_\ell + y$ *is a strictly contractive extension of* ψ_ℓ.

If G in (1.4) is strictly contractive, then G is said to be a *strictly contractive extension* of Φ (in (1.3)). Note that x in (1.8) may be rewritten as

$$x = x_1^{-1}(\psi_\ell + y)x_2^{-1} - \varphi_\ell - \omega. \tag{1.9}$$

Thus Theorem 1.1 tells us that the general form of the $(1,2)$-entry in a strictly contractive extension of Φ is equal to $x_1^{-1}\varphi x_2^{-1} - \omega$, where φ is an arbitrary strictly contractive extension of ψ_ℓ. In this way one may derive a linear fractional description of all strictly contractive extensions of Φ by substituting in (1.9) the linear fractional representation of all solutions $\varphi = \psi_\ell + y$ of the corresponding Nehari problem.

We call $G \in \mathcal{B}$ a *triangular extension* of Φ in (1.3) if G is a strictly contractive extension of ψ and the (1,2) entry of $G(E - G^*G)^{-1}$ is in $\mathcal{N}_\ell$.

THEOREM 1.2. *Let Φ be as in (1.3). Then Φ has a triangular extension*

(1.10) $$G = \begin{pmatrix} \varphi_{11} & \varphi_\ell + x \\ \varphi_{21} & \varphi_{22} \end{pmatrix} \in \mathcal{B} = \mathcal{N}^{2\times 2}$$

if and only if (1.5), (1.6) *hold and the equation*

(1.11) $$z - P_{\mathcal{N}_\ell}\{\psi_\ell P_{\mathcal{N}_u}(\psi_\ell^* z)\} = e$$

has a solution $z \in \mathcal{N}_\ell$ such that $z^{-1} \in \mathcal{N}_\ell$ and $P_{\mathcal{N}_d} z$ is positive definite in $\mathcal{N}_d$. Here

(1.12) $$\psi_\ell = P_{\mathcal{N}_\ell}\{x_1(\varphi_\ell + \omega)x_2\}$$

with x_1, x_2 and ω as in (1.7a,b,c). *Furthermore, in that case Φ has a unique triangular solution G which one obtains by inserting*

(1.13) $$x = -\varphi_\ell - \omega + x_1^{-1}(z^{-1})^*[P_{\mathcal{N}_u}(\psi_\ell^* z)]^* x_2^{-1}$$

into (1.10).

Theorem 1.2 tells us that G in (1.10) is the triangular extension of Φ if and only if $g := x_1(x + \varphi_\ell + \omega)x_2$ is the triangular extension of ψ_ℓ, i.e., g is a strictly contractive extension of ψ_ℓ and $g(e - g^*g)^{-1} \in \mathcal{N}_\ell$ (cf., [11], Theorem I.2.1). From the proof of Theorem 1.2 we shall derive a left spectral factorization for $E - GG^*$, when G is the triangular extension of Φ.

2. PROOFS OF THE MAIN THEOREMS

Throughout this section $\mathcal{N}$ is the algebra with triangular structure (1.1) introduced in the previous section, and $\mathcal{N}$ is assumed to satisfy axioms (E0) and (E1). As before $\mathcal{B} = \mathcal{N}^{2\times 2}$. We shall need the direct sum decomposition

(2.1) $$\mathcal{B} = \mathcal{B}_-^0 \dot{+} \mathcal{B}_d \dot{+} \mathcal{B}_+^0,$$

where

$$\mathcal{B}_-^0 = \left\{ \begin{pmatrix} a & 0 \\ c & d \end{pmatrix} \in \mathcal{B} \mid a, d \in \mathcal{N}_\ell^0 \right\},$$

$$\mathcal{B}_d = \left\{ \begin{pmatrix} a & 0 \\ 0 & d \end{pmatrix} \in \mathcal{B} \mid a, d \in \mathcal{N}_d \right\},$$

$$\mathcal{B}_+^0 = \left\{ \begin{pmatrix} a & b \\ 0 & d \end{pmatrix} \in \mathcal{B} \mid a, d \in \mathcal{N}_u^0 \right\}.$$

Note that (2.1) is a triangular structure on $\mathcal{B}$. It will be convenient to consider also the decomposition:

$$\mathcal{B} = \mathcal{B}_\ell \dot{+} \mathcal{B}_u, \tag{2.2}$$

where

$$\mathcal{B}_\ell = \left\{ \begin{pmatrix} a & b \\ c & d \end{pmatrix} \in \mathcal{B} \mid b \in \mathcal{N}_\ell \right\},$$

$$\mathcal{B}_u = \left\{ \begin{pmatrix} 0 & b \\ 0 & 0 \end{pmatrix} \in \mathcal{B} \mid b \in \mathcal{N}_u^0 \right\}.$$

For later purposes we mention the following facts:

$$\mathcal{B}_u \mathcal{B}_+^0 \subset \mathcal{B}_u, \quad \mathcal{B}_+^0 \mathcal{B}_u \subset \mathcal{B}_u, \tag{2.3}$$

$$\mathcal{B}_\ell \mathcal{B}_-^0 \subset \mathcal{B}_\ell, \quad \mathcal{B}_-^0 \mathcal{B}_\ell \subset \mathcal{B}_\ell, \tag{2.4}$$

$$\mathcal{B}_u \mathcal{B}_d \subset \mathcal{B}_u, \quad \mathcal{B}_d \mathcal{B}_u \subset \mathcal{B}_u, \tag{2.5}$$

$$\mathcal{B}_\ell \mathcal{B}_d \subset \mathcal{B}_\ell, \quad \mathcal{B}_\ell \mathcal{B}_d \subset \mathcal{B}_\ell, \tag{2.6}$$

Now, let $\mathcal{M} = \mathcal{B}^{2\times 2}$, and consider

$$\mathcal{M}_1 = \begin{pmatrix} 0 & \mathcal{B}_u \\ 0 & 0 \end{pmatrix} = \left\{ \begin{pmatrix} 0 & 0 & 0 & b \\ 0 & 0 & 0 & 0 \\ 0 & 0 & 0 & 0 \\ 0 & 0 & 0 & 0 \end{pmatrix} \mid b \in \mathcal{N}_u^0 \right\},$$

$$\mathcal{M}_2^0 = \begin{pmatrix} \mathcal{B}_+^0 & \mathcal{B}_\ell \\ 0 & \mathcal{B}_+^0 \end{pmatrix} = \left\{ \begin{pmatrix} a_{11} & * & * & b \\ 0 & a_{22} & * & * \\ 0 & 0 & a_{33} & * \\ 0 & 0 & 0 & a_{44} \end{pmatrix} \mid b \in \mathcal{N}_\ell, \quad a_{ii} \in \mathcal{N}_u^0, \quad i = 1, \dots, 4 \right\},$$

$$\mathcal{M}_d = \begin{pmatrix} \mathcal{B}_d & 0 \\ 0 & \mathcal{B}_d \end{pmatrix} = \left\{ \begin{pmatrix} a_{11} & 0 & 0 & 0 \\ 0 & a_{22} & 0 & 0 \\ 0 & 0 & a_{33} & 0 \\ 0 & 0 & 0 & a_{44} \end{pmatrix} \mid a_{ii} \in \mathcal{N}_d, \quad i = 1, \dots, 4 \right\},$$

$$\mathcal{M}_3^0 = \begin{pmatrix} \mathcal{B}_-^0 & 0 \\ \mathcal{B}_\ell^* & \mathcal{B}_-^0 \end{pmatrix} = \left\{ \begin{pmatrix} a_{11} & 0 & 0 & 0 \\ * & a_{22} & 0 & 0 \\ * & * & a_{33} & 0 \\ c & * & * & a_{44} \end{pmatrix} \mid c \in \mathcal{N}_u, \quad a_{ii} \in \mathcal{N}_\ell^0, \quad i = 1, \dots, 4 \right\},$$

$$\mathcal{M}_4 = \begin{pmatrix} 0 & 0 \\ \mathcal{B}_u^* & 0 \end{pmatrix} = \left\{ \begin{pmatrix} 0 & 0 & 0 & 0 \\ 0 & 0 & 0 & 0 \\ 0 & 0 & 0 & 0 \\ c & 0 & 0 & 0 \end{pmatrix} \mid c \in \mathcal{N}_\ell^0 \right\},$$

where the $*$'s denote arbitrary elements of $\mathcal{N}$. We have

$$\mathcal{M} = \mathcal{M}_1 \dot{+} \mathcal{M}_2^0 \dot{+} \mathcal{M}_d \dot{+} \mathcal{M}_3^0 \dot{+} \mathcal{M}_4, \tag{2.7}$$

and by using (2.3) - (2.6) it is straightforward to check that $\mathcal{M}$ with the decomposition (2.7) is an algebra with band structure, that is (see [10] -[13]),

$$\mathcal{M}_1 = \mathcal{M}_4^*, \quad \mathcal{M}_2^0 = (\mathcal{M}_3^0)^*, \quad \mathcal{M}_d = \mathcal{M}_d^*, \tag{2.8}$$

and the following multiplication table holds

	$\mathcal{M}_1$	$\mathcal{M}_2^0$	$\mathcal{M}_d$	$\mathcal{M}_3^0$	$\mathcal{M}_4$
$\mathcal{M}_1$	$\mathcal{M}_1$	$\mathcal{M}_1$	$\mathcal{M}_1$	$\mathcal{M}_+^0$	$\mathcal{M}$
$\mathcal{M}_2^0$	$\mathcal{M}_1$	$\mathcal{M}_+^0$	$\mathcal{M}_2^0$	$\mathcal{M}_c$	$\mathcal{M}_-^0$
$\mathcal{M}_d$	$\mathcal{M}_1$	$\mathcal{M}_2^0$	$\mathcal{M}_d$	$\mathcal{M}_3^0$	$\mathcal{M}_4$
$\mathcal{M}_3^0$	$\mathcal{M}_+^0$	$\mathcal{M}_c$	$\mathcal{M}_3^0$	$\mathcal{M}_-^0$	$\mathcal{M}_4$
$\mathcal{M}_4$	$\mathcal{M}$	$\mathcal{M}_-^0$	$\mathcal{M}_4$	$\mathcal{M}_4$	$\mathcal{M}_4$.

where

$$\mathcal{M}_+^0 := \mathcal{M}_1 \dot{+} \mathcal{M}_2^0, \quad \mathcal{M}_-^0 := \mathcal{M}_3^0 \dot{+} \mathcal{M}_4, \quad \mathcal{M}_c := \mathcal{M}_2^0 \dot{+} \mathcal{M}_d \dot{+} \mathcal{M}_3^0.$$

The involution * in (2.8) is the natural involution on $\mathcal{M}$ induced by the involution on $\mathcal{N}$.

In the sequel we use freely the notation introduced in the first three paragraphs of Section I.1 of [12] for the present $\mathcal{M}$. Note that

$$\mathcal{M}_+ = \begin{pmatrix} \mathcal{B}_+ & \mathcal{B} \\ 0 & \mathcal{B}_+ \end{pmatrix}, \quad \mathcal{M}_- = \begin{pmatrix} \mathcal{B}_- & 0 \\ \mathcal{B} & \mathcal{B}_- \end{pmatrix}, \quad \mathcal{M}_c = \begin{pmatrix} \mathcal{B} & \mathcal{B}_\ell \\ \mathcal{B}_\ell^* & \mathcal{B} \end{pmatrix},$$

where $\mathcal{B}_+ = \mathcal{B}_+^0 + \mathcal{B}_d$ and $\mathcal{B}_- = \mathcal{B}_-^0 + \mathcal{B}_d$. Let Φ be as in (1.3), and put

$$K = \begin{pmatrix} e & 0 & \varphi_{11} & \varphi_\ell \\ 0 & e & \varphi_{21} & \varphi_{22} \\ \varphi_{11}^* & \varphi_{21}^* & e & 0 \\ \varphi_\ell^* & \varphi_{22}^* & 0 & e \end{pmatrix} = \begin{pmatrix} E & \Phi \\ \Phi^* & E \end{pmatrix}.$$

We have $K \in \mathcal{M}_c$. Now let $G \in \mathcal{B} = \mathcal{N}^{2\times 2}$ be an extension of Φ, i.e., $G - \Phi \in \mathcal{B}_u$. Put

$$K_G = \begin{pmatrix} E & G \\ G^* & E \end{pmatrix}.$$

Then K_G is an extension of K, that is, $K_G - K$ belongs to $\mathcal{M}_1 \dot{+} \mathcal{M}_4$.

LEMMA 2.1. *Let $G \in \mathcal{B}$. Then G is a strictly contractive (resp., triangular) extension of Φ if and only if K_G is a positive (resp., band) extension of K.*

PROOF. Use axiom (E0) and the following identities:

$$K_G = \begin{pmatrix} E & 0 \\ G^* & E \end{pmatrix} \begin{pmatrix} E & 0 \\ 0 & E - G^*G \end{pmatrix} \begin{pmatrix} E & G \\ 0 & E \end{pmatrix}, \tag{2.9}$$

$$K_G^{-1} = \begin{pmatrix} * & -G(E - G^*G)^{-1} \\ * & * \end{pmatrix}. \tag{2.10}$$

A straightforward reasoning gives the desired results (see [11], Section I.2 for further details). ■

Note that any extension of K is of the form K_G for some $G \in \mathcal{B}$.

LEMMA 2.2. *Let $G \in \mathcal{B}$. Then $E - GG^*$ admits a left spectral factorization relative to* (2.1) *if and only if K_G admits a left spectral factorization relative to* (2.7).

PROOF. Assume $I - GG^* = H^*H$ with $H^{\pm 1} \in \mathcal{B}_- = \mathcal{B}_-^0 \dot{+} \mathcal{B}_d$. Since

$$K_G = \begin{pmatrix} E & G \\ 0 & E \end{pmatrix} \begin{pmatrix} E - GG^* & 0 \\ 0 & E \end{pmatrix} \begin{pmatrix} E & 0 \\ G^* & E \end{pmatrix}, \tag{2.11}$$

we see that K_G factors as $K_G = L^*L$ with $L = \begin{pmatrix} H & 0 \\ G^* & E \end{pmatrix}$. Now

$$\begin{pmatrix} H & 0 \\ G^* & E \end{pmatrix} \in \mathcal{M}_-, \quad \begin{pmatrix} H & 0 \\ G^* & E \end{pmatrix}^{-1} = \begin{pmatrix} H^{-1} & 0 \\ -G^*H^{-1} & E \end{pmatrix} \in \mathcal{M}_-.$$

Thus $K_G = L^*L$ is a left spectral factorization of K_G relative to (2.7).

Next, let $K_G = L^*L$ be an arbitrary left spectral factorization of K relative to (2.7). Thus

$$L = \begin{pmatrix} A & 0 \\ C & D \end{pmatrix}, \qquad L^{-1} = \begin{pmatrix} A^\times & 0 \\ C^\times & D^\times \end{pmatrix}$$

with $A, A^\times, D$ and $D^\times$ in $\mathcal{B}_-$. Note that $DD^\times = D^\times D = E$. Thus D is invertible (and $D^{\pm 1} \in \mathcal{B}_-$). From

$$\begin{pmatrix} E & G \\ G^* & E \end{pmatrix} = \begin{pmatrix} A^* & C^* \\ 0 & D^* \end{pmatrix} \begin{pmatrix} A & 0 \\ C & D \end{pmatrix}, \tag{2.12}$$

we see that $G^* = D^*C$ and $D^*D = E$. Because of the latter identity, we have $D^* = D^{-1}$, and thus $C - DG^* = 0$. Now use (2.12), and rewrite (2.11) as

$$\begin{pmatrix} E-GG^* & 0 \\ 0 & E \end{pmatrix} = \begin{pmatrix} E & -G \\ 0 & E \end{pmatrix} \begin{pmatrix} A^* & C^* \\ 0 & D^* \end{pmatrix} \begin{pmatrix} A & 0 \\ C & D \end{pmatrix} \begin{pmatrix} E & 0 \\ -G^* & E \end{pmatrix}$$
$$= \begin{pmatrix} A & 0 \\ 0 & D \end{pmatrix}^* \begin{pmatrix} A & 0 \\ 0 & D \end{pmatrix}.$$

So $E - GG^* = A^*A$, where $A \in \mathcal{B}_-$. Finally, note that

$$\begin{pmatrix} A^\times & 0 \\ C^\times & D^\times \end{pmatrix} = \begin{pmatrix} A & 0 \\ C & D \end{pmatrix}^{-1} \begin{pmatrix} A^{-1} & 0 \\ * & * \end{pmatrix}.$$

Hence $A^{-1} = A^\times \in \mathcal{B}_-$. So $E - GG^*$ has the desired spectral factorization. ■

Consider the matrices

$$\Lambda_1 = \begin{pmatrix} e & 0 & \varphi_{11} \\ 0 & e & \varphi_{21} \\ \varphi_{11}^* & \varphi_{21}^* & e \end{pmatrix}, \quad \Lambda_2 = \begin{pmatrix} e & \varphi_{21} & \varphi_{22} \\ \varphi_{21}^* & e & 0 \\ \varphi_{22}^* & 0 & e \end{pmatrix}, \tag{2.13}$$

and let Δ and δ be defined by (1.5) and (1.6), respectively.

LEMMA 2.3. *We have $\Lambda_1 > 0$ (resp., $\Lambda_2 > 0$) if and only if $\Delta > 0$ (resp., $\delta > 0$). Furthermore,*

$$\Lambda_1^{-1} = \begin{pmatrix} e + \varphi_{11}\Delta^{-1}\varphi_{11}^* & \varphi_{11}\Delta^{-1}\varphi_{21}^* & -\varphi_{11}\Delta^{-1} \\ \varphi_{21}\Delta^{-1}\varphi_{11}^* & e + \varphi_{21}\Delta^{-1}\varphi_{21}^* & -\varphi_{21}\Delta^{-1} \\ -\Delta^{-1}\varphi_{11}^* & -\Delta^{-1}\varphi_{21}^* & \Delta^{-1} \end{pmatrix}, \tag{2.14}$$

$$\Lambda_2^{-1} = \begin{pmatrix} \delta^{-1} & -\delta^{-1}\varphi_{21} & -\delta^{-1}\varphi_{22} \\ -\varphi_{21}^*\delta^{-1} & e + \varphi_{21}^*\delta^{-1}\varphi_{21} & \varphi_{21}^*\delta^{-1}\varphi_{22} \\ -\varphi_{22}^*\delta^{-1} & \varphi_{22}^*\delta^{-1}\varphi_{21} & e + \varphi_{22}^*\delta^{-1}\varphi_{22} \end{pmatrix}, \tag{2.15}$$

whenever Δ or δ is invertible.

PROOF. To prove the statements concerning Λ_1, use axiom (E0) and a factorization of the type (2.9) with $G = \begin{pmatrix} \varphi_{11} \\ \varphi_{21} \end{pmatrix}$. For Λ_2, replace (2.9) by (2.11), and apply with $G = (\,\varphi_{21} \quad \varphi_{22}\,)$. ■

PROPOSITION 2.4. *There exists*

$$\widetilde{B} = \begin{pmatrix} e & 0 & \varphi_{11} & b \\ 0 & e & \varphi_{21} & \varphi_{22} \\ \varphi_{11}^* & \varphi_{21}^* & e & 0 \\ b^* & \varphi_{22}^* & 0 & e \end{pmatrix} \in \mathcal{B}^{2\times 2} \tag{2.16}$$

such that $\widetilde{B} > 0$ and the $(1,4)$-entry of $\widetilde{B}^{-1}$ is zero if and only if Δ and δ are positive definite. In that case there exists only one such $\widetilde{B}$ which one obtains by taking

$$b = -(e + \varphi_{11}\Delta^{-1}\varphi_{11}^*)^{-1}\varphi_{11}\Delta^{-1}\varphi_{21}^*\varphi_{22}. \tag{2.17}$$

PROOF. Assume $\widetilde{B}$ in (2.16) is positive definite. Then, by axiom (E0), the matrices Λ_1 and Λ_2 are positive definite. So, by Lemma 2.3, we have $\Delta > 0$ and $\delta > 0$.

Conversely, assume that Δ and δ are positive definite. We have to solve a band extension problem in the algebra $\mathcal{M}$ of 4×4 matrices with entries in $\mathcal{N}$ for the band given by

$$\begin{pmatrix} e & 0 & \varphi_{11} & ? \\ 0 & e & \varphi_{21} & \varphi_{22} \\ \varphi_{11}^* & \varphi_{21}^* & e & 0 \\ ? & \varphi_{22}^* & 0 & e \end{pmatrix}.$$

To do this we apply the band method. Consider the following subspaces

$$\mathcal{L}_1 = \{(a_{ij})_{i,j=1}^4 \in \mathcal{N}^{4\times 4} \mid a_{ij} = 0 \text{ for } j - i \leq 2\},$$

$$\mathcal{L}_2^0 = \{(a_{ij})_{i,j=1}^4 \in \mathcal{N}^{4\times 4} \mid a_{ij} = 0 \text{ for } j - i > 2 \text{ or } j - i \leq 0\},$$

$$\mathcal{L}_d = \{(a_{ij})_{i,j=1}^4 \in \mathcal{N}^{4\times 4} \mid a_{ij} = 0 \text{ for } j \neq i\},$$

$$\mathcal{L}_3^0 = \{(a_{ij})_{i,j=1}^4 \in \mathcal{N}^{4\times 4} \mid a_{ij} = 0 \text{ for } j - i \geq 0 \text{ or } j - i < -2\},$$

$$\mathcal{L}_4^0 = \{(a_{ij})_{i,j=1}^4 \in \mathcal{N}^{4\times 4} \mid a_{ij} = 0 \text{ for } j - i \geq -2\}.$$

Then

$$\mathcal{M} = \mathcal{L}_1 \dot{+} \mathcal{L}_2^0 \dot{+} \mathcal{L}_d \dot{+} \mathcal{L}_3^0 \dot{+} \mathcal{L}_4, \tag{2.18}$$

and (2.18) defines a band structure on $\mathcal{M}$ (which is different from the one associated with (2.7)). Note that

$$K_0 = \begin{pmatrix} e & 0 & \varphi_{11} & 0 \\ 0 & e & \varphi_{21} & \varphi_{22} \\ \varphi_{11}^* & \varphi_{21}^* & e & 0 \\ 0 & \varphi_{22}^* & 0 & e \end{pmatrix} \in \mathcal{L}_c := \mathcal{L}_2^0 \dot{+} \mathcal{L}_d \dot{+} \mathcal{L}_3^0.$$

We shall apply Theorems I.1.3 and I.1.4 in [10] for the band structure defined by (2.18). Since $\Delta > 0$, we know from Lemma 2.3 that $\Lambda_1 > 0$, and hence (use Axiom (E0)) we may conclude that $e + \varphi_{11}\Delta^{-1}\varphi_{11}^*$ is positive definite in $\mathcal{N}$. Now, apply axiom (E1) to show that we may choose x_1 as in (1.7a). We may also factor δ^{-1} as

$$\delta^{-1} = u^* u, \tag{2.20}$$

where u is invertible in $\mathcal{N}$. Put

$$\widetilde{Y} = \begin{pmatrix} e + \varphi_{11}\Delta^{-1}\varphi_{11}^* & 0 & 0 & 0 \\ \varphi_{21}\Delta^{-1}\varphi_{11}^* & \delta^{-1} & 0 & 0 \\ -\Delta^{-1}\varphi_{11}^* & -\varphi_{21}^*\delta^{-1} & e & 0 \\ 0 & -\varphi_{22}^*\delta^{-1} & 0 & e \end{pmatrix}, \quad \widetilde{D} = \begin{pmatrix} x_1 & 0 & 0 & 0 \\ 0 & u & 0 & 0 \\ 0 & 0 & e & 0 \\ 0 & 0 & 0 & e \end{pmatrix}.$$

Then $\widetilde{Y} \in \mathcal{L}_3 := \mathcal{L}_3^0 \dot{+} \mathcal{L}_d$ and $\widetilde{Y}^{-1} \subset \mathcal{L}_- := \mathcal{L}_4 \dot{+} \mathcal{L}_3^0 \dot{+} \mathcal{L}_d$. Furthermore, the diagonal of $\widetilde{Y}$ is equal to $\widetilde{D}^*\widetilde{D}$, and hence positive definite in $\mathcal{L}_d$. So we can apply Theorem I.1.3 in [10] to show that

$$\widetilde{B} := (\widetilde{Y}^{-1})^* \widetilde{D}^* \widetilde{D} \widetilde{Y}^{-1} \tag{2.21}$$

is a band extension of K_0 (in (2.19)) relative to the band structure (2.18). Since $\widetilde{Y}$ is lower triangular, it follows that the (1,4)-entry of $\widetilde{B}$ is equal to the adjoint of $(\widetilde{Y}^{-1})_{41}$, which one computes to be the element given by the right hand side of (2.17).

To prove the uniqueness of $\widetilde{B}$ we first note that axioms (E0) and (E1) imply that the positive definite elements in $\mathcal{N}^{4\times 4}$ admit left and right spectral factorizations relative to (2.18) (cf., the proof of Lemma 2.5 at the end of this section). But then we can apply Theorem I.1.4 in [10] to show that K_0 has a unique band extension relative to (2.18). ■

The element b in (2.17) is also given by the following formula

$$b = -\varphi_{11}\varphi_{21}^*\delta^{-1}\varphi_{22}(e + \varphi_{22}^*\delta^{-1}\varphi_{22})^{-1}, \tag{2.22}$$

which one may prove by direct checking.

PROOF OF THEOREM 1.1. We divide the proof into five parts.

Part (a). Let G be a strictly contractive extension of Φ. Then $K_G > 0$ by Lemma 2.1. Thus, by axiom (E0) the matrices Λ_1 and Λ_2 are positive definite. Therefore, by Lemma 2.3, we have $\Delta > 0$ and $\delta > 0$.

Part (b). Assume $\Delta > 0$ and $\delta > 0$. By Lemma 2.3, the block matrices Λ_1 and Λ_2 are positive definite in $\mathcal{N}^{3\times 3}$. It follows that

$$\begin{pmatrix} \Lambda_1^{-1} & 0 \\ 0 & e \end{pmatrix} > 0, \qquad \begin{pmatrix} e & 0 \\ 0 & \Lambda_2^{-1} \end{pmatrix} > 0.$$

Thus, by axiom (E0), the (1,1)-entry of Λ_1^{-1} and the (3,3)-entry of Λ_2^{-1} are positive definite. But then we can apply axiom (E1) to show the existence of the factorizations (1.7a) and (1.7b).

Part (c). Assume $\Delta > 0$ and $\delta > 0$, and let G be as in (1.4) with $x \in \mathcal{N}_u^0$. We may apply Proposition 2.4. Let $\widetilde{B}$ be as in (2.16) with b as in (2.17). Recall that $\widetilde{B}$ is also given by (2.21), and note that $b = -\omega$, where ω is given by (1.7c). It follows that

$$K_G = M_1 + \widetilde{B} + M_1^*,$$

where

$$M_1 = \begin{pmatrix} 0 & 0 & 0 & \ell + x + \omega \\ 0 & 0 & 0 & 0 \\ 0 & 0 & 0 & 0 \\ 0 & 0 & 0 & 0 \end{pmatrix}.$$

By (2.21) we have $\widetilde{B} = (H^{-1})^* H^{-1}$ with

$$H = \widetilde{Y}\widetilde{D}^{-1} = \begin{pmatrix} x_1^* & 0 & 0 & 0 \\ \varphi_{21}\Delta^{-1}\varphi_{11}^* x_1^{-1} & u^* & 0 & 0 \\ -\Delta^{-1}\varphi_{11}^* x_1^{-1} & -\varphi_{21}^* u^* & e & 0 \\ 0 & -\varphi_{22}^* u^* & 0 & e \end{pmatrix},$$

where u is as in (2.20). We see that

$$H^* K_G H = H^* M_1 H + \widetilde{E} + H M_1^* H^*,$$

where $\widetilde{E}$ is the unit of $\mathcal{B}^{2\times 2}$ and

$$H^* M_1 H = \begin{pmatrix} 0 & -x_1\psi\varphi_{22}^* u^* & 0 & x_1\psi \\ 0 & 0 & 0 & 0 \\ 0 & 0 & 0 & 0 \\ 0 & 0 & 0 & 0 \end{pmatrix}$$

with $\psi = \varphi_\ell + x + \omega$. In this way we derive that G is a strictly contractive extension of Φ if and only if

$$(2.23) \qquad \begin{pmatrix} e & -x_1\psi\varphi_{22}^* u^* & 0 & x_1\psi_1 \\ -u\varphi_{22}\psi^* x_1^* & e & 0 & 0 \\ 0 & 0 & e & 0 \\ \psi^* x_1^* & 0 & 0 & e \end{pmatrix}$$

is positive definite. Note that

$$e - \begin{pmatrix} -x_1\psi\varphi_{22}^* u^* & x_1\psi \end{pmatrix} \begin{pmatrix} -u\varphi_{22}\psi^* x_1^* \\ \psi^* x_1^* \end{pmatrix} = e - x_1\psi(e + \varphi_{22}^*\delta^{-1}\varphi_{22})\psi^* x_1^*$$
$$= e - (x_1\psi x_2)(x_1\psi x_2)^*,$$

where we use (1.7b). It follows that G is a strictly contractive extension of Φ if and only if

$$e - (x_1\psi x_2)(x_1\psi x_2)^* > 0. \tag{2.24}$$

Part (d). Let G be a strictly contractive extension of Φ. Put $\psi_\ell = P_{\mathcal{N}_\ell}\{x_1(\varphi_\ell + \omega)x_2\}$. We have $x_1 x x_2 \in \mathcal{N}_u$, because $x \in \mathcal{N}_u^0$ and x_1, x_2 in $\mathcal{N}_u$. Thus

$$x_1\psi x_2 = \psi_\ell + P_{\mathcal{N}_u^0}\{x_1(\varphi_\ell + \omega)x_2\} + x_1 x x_2. \tag{2.25}$$

From (2.25) we see that $\psi_\ell + y$, where

$$y = P_{\mathcal{N}_u^0}\{x_1(\varphi_\ell + \omega)x_2\} + x_1 x x_2 \tag{2.26}$$

is a strictly contractive extension of ψ_ℓ. Thus ψ_ℓ has a strictly contractive extension in $\mathcal{N}$ and x has the desired representation.

Part (e). Let $\psi_\ell + y$ be a strictly contractive extension of ψ_ℓ. Define x by (1.8). Since x_1^{-1} and x_2^{-1} are in $\mathcal{N}_u$, we have $x \in \mathcal{N}_u$. Let G be as in (1.4) with this choice of x, and put $\psi = \varphi_\ell + x + \omega$. Then (2.25) and (2.26) hold, and hence we have also (2.24). Thus, by the result of part (c), G is a strictly contractive extension of Φ. ■

PROOF OF THEOREM 1.2. If Φ has a triangular extension, then, by Theorem 1.1, we have $\Delta > 0$ and $\delta > 0$. So in what follows we may assume that both Δ and δ are positive definite. Furthermore, we assume that x_1, x_2 and ω are as in (1.7a,b,c). We split the proof into three parts. Sometimes we write a^{-*} for the adjoint of a^{-1}.

Part (a). By Lemma 2.1 it suffices to look for a band extension of

$$K = \begin{pmatrix} E & \Phi \\ \Phi^* & E \end{pmatrix}. \tag{2.27}$$

Axioms (E0) and (E1) imply (see Lemma 2.5 below) that the positive definite elements in $\mathcal{M}$ have a left and right spectral factorization relative to the band structure (2.7). Hence (see Theorem I.1.1 in [11]) the band extension of K (assuming it exists) is unique.

Part (b). In this part we apply Theorem I.1.1 in [11] to construct a band extension for K assuming (1.11) has the desired solution. We seek

$$Y = \begin{pmatrix} a_{11} & 0 & 0 & 0 \\ a_{21} & a_{22} & 0 & 0 \\ c_{11} & c_{12} & d_{11} & 0 \\ c_u & c_{22} & d_{21} & d_{22} \end{pmatrix} \tag{2.28}$$

such that

$$P_3(KY) = \begin{pmatrix} e & 0 & 0 & 0 \\ 0 & e & 0 & 0 \\ 0 & 0 & e & 0 \\ 0 & 0 & 0 & e \end{pmatrix} \tag{2.29}$$

and

(a) $a_{ii} \in \mathcal{N}_\ell, \quad d_{ii} \in \mathcal{N}_\ell \ (i = 1, 2), \quad c_u \in \mathcal{N}_u$

(b) a_{ii} and d_{ii} are positive definite in $\mathcal{N}_d$ $(i = 1, 2)$;

(c) $a_{ii}^{-1} \in \mathcal{N}_\ell, \quad d_{ii}^{-1} \in \mathcal{N}_\ell \ (i = 1, 2)$.

One computes that (2.29) is equivalent to

$$\begin{pmatrix} d_{11} & 0 \\ d_{21} & d_{22} \end{pmatrix} = \begin{pmatrix} e & 0 \\ 0 & e \end{pmatrix}$$

and the following seven equations:

$$P_{\mathcal{N}_\ell}(a_{11} + \varphi_{11}c_{11} + \varphi_\ell c_u) = e,$$

$$P_{\mathcal{N}_\ell}(a_{22} + \varphi_{21}c_{12} + \varphi_{22}c_{22}) = e,$$

$$a_{21} + \varphi_{21}c_{11} + \varphi_{22}c_u = 0,$$

$$\varphi_{11}^* a_{11} + \varphi_{21}^* a_{21} + c_{11} = 0,$$

$$\varphi_{21}^* a_{22} + c_{12} = 0,$$

$$P_{\mathcal{N}_u}(\varphi_\ell^* a_{11} + \varphi_{22}^* a_{21} + c_u) = 0,$$

$$\varphi_{22}^* a_{22} + c_{22} = 0.$$

These seven equations may be rewritten in the following way:

$$\begin{pmatrix} e & \varphi_{21} & \varphi_{22} \\ \varphi_{21}^* & e & 0 \\ \varphi_{22}^* & 0 & e \end{pmatrix} \begin{pmatrix} a_{22} \\ c_{12} \\ c_{22} \end{pmatrix} = \begin{pmatrix} e \\ 0 \\ 0 \end{pmatrix} + \begin{pmatrix} r_u^0 \\ 0 \\ 0 \end{pmatrix}, \tag{2.30a}$$

$$\text{(2.30b)}\qquad \begin{pmatrix} e & 0 & \varphi_{11} & \varphi_\ell \\ 0 & e & \varphi_{21} & \varphi_{22} \\ \varphi_{11}^* & \varphi_{21}^* & e & 0 \\ \varphi_\ell^* & \varphi_{22}^* & 0 & e \end{pmatrix} \begin{pmatrix} a_{11} \\ a_{21} \\ c_{11} \\ c_u \end{pmatrix} = \begin{pmatrix} e \\ 0 \\ 0 \\ 0 \end{pmatrix} + \begin{pmatrix} n_u^0 \\ 0 \\ 0 \\ n_\ell^0 \end{pmatrix},$$

where $r_u^0, n_u^0 \in \mathcal{N}_u^0$ and $n_\ell^0 \in \mathcal{N}_\ell^0$ are free to choose. From (2.15) we see that

$$\text{(2.31)}\qquad \begin{pmatrix} a_{22} \\ c_{12} \\ c_{22} \end{pmatrix} = \begin{pmatrix} e \\ -\varphi_{21}^* \\ -\varphi_{22}^* \end{pmatrix} (\delta^{-1} + \delta^{-1} r_u^0).$$

Now factor δ^{-1} as $\delta^{-1} = u^* u$ with $u^{\pm 1} \in \mathcal{N}_u$, which is possible because of axiom (E1). Then

$$\text{(2.32)}\qquad u^{-*} a_{22} = u + u r_u^0.$$

Note that $u r_u^0 \in \mathcal{N}_u^0$. We seek $a_{22} \in \mathcal{N}_\ell$. So $u^{-*} a_{22} \in \mathcal{N}_\ell$. It follows that we must take $r_u^0 \in \mathcal{N}_u^0$ so that $P_{\mathcal{N}_u^0} u = -u r_u^0$. In other words,

$$\text{(2.33)}\qquad r_u^0 = -u^{-1} P_{\mathcal{N}_u^0} u.$$

Write $u = u_d + u^0$ with $u_d \in \mathcal{N}_d$ and $u_0 \in \mathcal{N}_u^0$. Then

$$\text{(2.34)}\qquad a_{22} = u^* u - u^* P_{\mathcal{N}_u^0} u = u^* u_d.$$

Since $u^{\pm 1} \in \mathcal{N}_u$, we conclude that $a_{22}^{\pm 1} \in \mathcal{N}_\ell$ and $a_{22} > 0$ in $\mathcal{N}$. So we found a_{22}, and from (2.31) we see that

$$\text{(2.35)}\qquad c_{12} = -\varphi_{21}^* a_{22}, \qquad c_{22} = -\varphi_{22}^* a_{22}.$$

Next, we solve (2.30b). By using (2.14) we see that (2.30b) may be rewritten as:

$$\text{(2.36a)}\qquad \varphi_\ell^* a_{11} + \varphi_{22}^* a_{21} + c_u = n_\ell^0,$$

$$\text{(2.36b)}\qquad \begin{pmatrix} a_{11} \\ a_{21} \\ c_{11} \end{pmatrix} + \begin{pmatrix} e + \varphi_{11}\Delta^{-1}\varphi_{11}^* \\ \varphi_{21}\Delta^{-1}\varphi_{11}^* \\ -\Delta^{-1}\varphi_{11}^* \end{pmatrix} \varphi_\ell c_u + \begin{pmatrix} \varphi_{11}\Delta^{-1}\varphi_{21}^* \\ e + \varphi_{21}\Delta^{-1}\varphi_{22}^* \\ -\Delta^{-1}\varphi_{21}^* \end{pmatrix} \varphi_{22} c_u = \begin{pmatrix} e + \varphi_{11}\Delta^{-1}\varphi_{11}^* \\ \varphi_{21}\Delta^{-1}\varphi_{11}^* \\ -\Delta^{-1}\varphi_{11}^* \end{pmatrix} (e + n_u^0).$$

The first row in (2.36b) gives:

$$a_{11} = -(e + \varphi_{11}\Delta^{-1}\varphi_{11}^*)\{(\varphi_\ell + \omega)c_u - (e + n_u^0)\},$$

where ω is defined by (1.7c). Now use (1.7a) to get

$$x_1^{-*}a_{11} = -x_1(\varphi_\ell + \omega)c_u + x_1(e + n_u^0). \tag{2.37}$$

The second row in (2.36b) and (2.37) yield:

$$\begin{aligned} a_{21} &= -\varphi_{21}\Delta^{-1}\varphi_{11}^*\varphi_\ell c_u - (e + \varphi_{21}\Delta^{-1}\varphi_{21}^*)\varphi_{22}c_u + \varphi_{21}\Delta^{-1}\varphi_{11}^*(e + n_u^0) \\ &= -\varphi_{21}\Delta^{-1}\varphi_{11}^*\varphi_\ell c_u - (e + \varphi_{21}\Delta^{-1}\varphi_{21}^*)\varphi_{22}c_u + \varphi_{21}\Delta^{-1}\varphi_{11}^*x_1^{-1}x_1^{-*}a_{11} \\ &\quad + \varphi_{21}\Delta^{-1}\varphi_{11}^*(\varphi_\ell + \omega)c_u \\ &= \varphi_{21}\Delta^{-1}\varphi_{11}^*\omega c_u - (e + \varphi_{21}\Delta^{-1}\varphi_{21}^*)\varphi_{22}c_u + \varphi_{21}\Delta^{-1}\varphi_{11}^*(e + \varphi_{11}\Delta^{-1}\varphi_{11}^*)^{-1}a_{11}. \end{aligned}$$

Thus

$$\varphi_{22}^*a_{21} = \omega^*a_{11} + \varphi_{22}^*\varphi_{21}\Delta^{-1}\varphi_{11}^*\omega c_u - \varphi_{22}^*(e + \varphi_{21}\Delta^{-1}\varphi_{21}^*)\varphi_{22}c_u.$$

It follows that

$$\varphi_\ell^*a_{11} + \varphi_{22}^*a_{21} + c_u = (\varphi_\ell^* + \omega^*)a_{11} + c_u + \alpha,$$

where

$$\alpha = \varphi_{22}^*\varphi_{21}\Delta^{-1}\varphi_{11}^*\omega c_u - \varphi_{22}^*(e + \varphi_{21}\Delta^{-1}\varphi_{21}^*)\varphi_{22}c_u.$$

Now use that

$$\Delta^{-1}\varphi_{11}^*(e + \varphi_{11}\Delta^{-1}\varphi_{11}^*)^{-1}\varphi_{11} = e - (\Delta + \varphi_{11}^*\varphi_{11})^{-1}\Delta.$$

It follows that

$$\begin{aligned} \alpha &= -\varphi_{22}^*\varphi_{22}c_u - \varphi_{22}^*\varphi_{21}\Delta^{-1}\varphi_{21}^*\varphi_{22}c_u \\ &\qquad + \varphi_{22}^*\varphi_{21}\Delta^{-1}\varphi_{11}^*(e + \varphi_{11}\Delta^{-1}\varphi_{11}^*)^{-1}\varphi_{11}\Delta^{-1}\varphi_{21}^*\varphi_{22}c_u \\ &= -\varphi_{22}^*\varphi_{22}c_u - \varphi_{22}^*\varphi_{21}(\Delta + \varphi_{11}^*\varphi_{11})^{-1}\varphi_{21}^*\varphi_{22}c_u \\ &= -\varphi_{22}^*\varphi_{22}c_u - \varphi_{22}^*[-e + (e - \varphi_{21}\varphi_{21}^*)^{-1}]\varphi_{22}c_u \\ &= -\varphi_{22}^*(e - \varphi_{21}\varphi_{21}^*)^{-1}\varphi_{22}c_u. \end{aligned}$$

Thus

$$\varphi_\ell^* a_{11} + \varphi_{22}^* a_{21} + c_u = (\varphi_\ell + \omega)^* a_{11} + c_u - \varphi_{22}^*(e - \varphi_{21}\varphi_{21}^*)^{-1}\varphi_{22} c_u$$
$$= (\varphi_\ell + \omega)^* a_{11} + [e - \varphi_{22}^*(e - \varphi_{21}\varphi_{21}^*)^{-1}\varphi_{22}] c_u$$
$$= (\varphi_\ell + \omega)^* a_{11} + [e + \varphi_{22}^*\delta^{-1}\varphi_{22}]^{-1} c_u.$$

Now use (1.7b) to show that

$$\varphi_\ell^* a_{11} + \varphi_{22}^* a_{21} + c_u \in \mathcal{N}_\ell^0 \iff \{x_2^*(\varphi_\ell + \omega)^* x_1^*\} x_1^{-*} a_{11} + x_2^{-1} c_u \in \mathcal{N}_\ell^0. \tag{2.38}$$

Now let us assume that (1.11) has the desired solution z. Put

$$a_{11} = x_1^* z d_1, \qquad d_1 = P_{\mathcal{N}_u} x_1. \tag{2.39}$$

Then $a_{11}^{\pm 1} \in \mathcal{N}_\ell$, because x_1^* and z have this property. Furthermore,

$$P_d(a_{11}) = d_1^*(P_d z) d_1, \tag{2.40}$$

and thus $P_d(a_{11})$ is positive definite in $\mathcal{N}_d$. Next define

$$c_u := -x_2[P_{\mathcal{N}_u}(\psi_\ell^* z)] d_1 \in \mathcal{N}_u, \tag{2.41}$$

$$a_{21} := \varphi_{21}\Delta^{-1}\varphi_{11}^*\omega c_u - (e + \varphi_{21}\Delta^{-1}\varphi_{21}^*)\varphi_{22} c_u + \varphi_{22}\Delta^{-1}\varphi_{11}^*(e + \varphi_{11}\Delta^{-1}\varphi_{11}^*)^{-1} a_{11}. \tag{2.42}$$

Then we know from (2.38) that (2.36a) holds for some $n_\ell^0 \in \mathcal{N}_\ell^0$. Furthermore, from (1.11) we see that

$$P_{\mathcal{N}_\ell}\{x_1^{-*} a_{11} d_1^{-1} + x_1(\varphi_\ell + \omega) x_3 (x_3^{-1} c_u d_1^{-1})\} = e, \tag{2.43}$$

and thus

$$s_u^0 := x_1^{-*} a_{11} + x_1(\varphi_\ell + \omega) c_u - x_1 \in \mathcal{N}_u^0.$$

So (2.37) holds true with $n_u^0 = x_1^{-1} s_u^0 \in \mathcal{N}_u^0$. Finally, define

$$c_{11} = \Delta^{-1}\varphi_{11}^*\varphi_\ell c_u + \Delta^{-1}\varphi_{21}^*\varphi_{22} c_u - \Delta^{-1}\varphi_{11}^*(e + n_u^0).$$

Then Y in (2.28) has the desired properties, and thus we can apply Theorem I.1.1 in [11] to show that

$$B = (Y^{-1})^*(P_d Y) Y^{-1}$$

is the desired band extension.

It remains to compute the $(1,4)$-entry of B. Note that

$$P_d(Y) = \begin{pmatrix} d_1^*(P_d z)d_1 & 0 & 0 & 0 \\ 0 & u_d^* u_d & 0 & 0 \\ 0 & 0 & e & 0 \\ 0 & 0 & 0 & e \end{pmatrix}. \tag{2.44}$$

It follows that

$$B_{14} = \left((Y^{-1})_{41}\right)^*.$$

So let us solve

$$\begin{pmatrix} a_{11} & 0 & 0 & 0 \\ a_{21} & a_{22} & 0 & 0 \\ c_{11} & c_{12} & e & 0 \\ c_u & c_{22} & 0 & e \end{pmatrix} \begin{pmatrix} t_1 \\ t_2 \\ t_3 \\ t_4 \end{pmatrix} = \begin{pmatrix} e \\ 0 \\ 0 \\ 0 \end{pmatrix}.$$

Then

$$\begin{aligned} (Y^{-1})_{41} = t_4 &= -c_u a_{11}^{-1} + c_{22} a_{22}^{-1} a_{21} a_{11}^{-1} \\ &= (-c_u - \varphi_{22}^* a_{21}) a_{11}^{-1} \\ &= -(\omega^* a_{11} + x_2^{-*} x_2^{-1} c_u) a_{11}^{-1} \\ &= -\omega^* + x_2^{-*}[P_{\mathcal{N}_u}(\psi_\ell^* z)] d_1 a_{11}^{-1} \\ &= -\omega^* + x_2^{-*}[P_{\mathcal{N}_u}(\psi_\ell^* z)] z^{-1} x_1^{-*}. \end{aligned}$$

Thus $B_{14} = \varphi_\ell + x$, where

$$x = -\varphi_\ell - \omega + x_1^{-1} z^{-*} [P_{\mathcal{N}_u}(\psi_\ell^* z)]^* x_2^{-1},$$

and hence x is of the form (1.13).

Part (c). In this part we assume that K has a band extension $B = K_G$. By Theorem I.1.1 in [11] this implies that the equation (2.29) has a solution Y with the properties (a), (b) and (c) mentioned directly after (2.29). Put $z = x_1^{-*} a_{11} d_1^{-1}$ as in (2.39). Then $z^{\pm 1} \in \mathcal{N}_\ell$, and $P_d(z) = d_1^{-*}(P_d a_{11}) d_1$, which is positive definite in $\mathcal{N}_d$. From (2.38) we see that

$$P_{\mathcal{N}_u}(\psi_\ell^* z) = -x_2^{-1} c_u d_1^{-1},$$

and (2.37) yields

$$z + P_{\mathcal{N}_\ell}(\psi_\ell x_2^{-1} c_u d_1^{-1}) = e.$$

By combining these two identities we see that (1.11) has a solution with the desired properties. ∎

LEMMA 2.5. *A positive definite element in* $\mathcal{B}^{2\times 2}$ *admits a left and a right spectral factorization relative to the band structure* (2.7).

PROOF. Let $A \in \mathcal{B}^{2\times 2}$ be positive definite. Then, by axiom (E0), the principal submatrices of A are positive definite. Hence we can factor A as $A = LDL^*$, where L and D are of the form

$$L = \begin{pmatrix} e & 0 & 0 & 0 \\ * & e & 0 & 0 \\ * & * & e & 0 \\ * & * & * & e \end{pmatrix}, \quad D = \begin{pmatrix} d_1 & 0 & 0 & 0 \\ 0 & d_2 & 0 & 0 \\ 0 & 0 & d_3 & 0 \\ 0 & 0 & 0 & d_4 \end{pmatrix}$$

with d_1, d_2, d_3 and d_4 positive definite in $\mathcal{N}$. By axiom (E1) we have

$$d_j = \widetilde{x}_j \widetilde{x}_j^*, \qquad \widetilde{x}_j^{\pm 1} \in \mathcal{N}_\ell.$$

Now,

$$M := L \begin{pmatrix} \widetilde{x}_1 & 0 & 0 & 0 \\ 0 & \widetilde{x}_2 & 0 & 0 \\ 0 & 0 & \widetilde{x}_3 & 0 \\ 0 & 0 & 0 & \widetilde{x}_4 \end{pmatrix} \in \mathcal{M}_-, \quad M^{-1} \in \mathcal{M}_-,$$

and $A = MM^*$ is a right spectral factorization relative to (2.7). ∎

REFERENCES

[1] J.C. Doyle, Synthesis of robust controllers, *Proc. IEEE Conf. Dec. Control*, 1983.

[2] J.C. Doyle, *Lecture notes in advances in multivariable control*, ONR/Honeywell Workshop, Minneapolis, 1984.

[3] H. Dym and I. Gohberg, A new class of contractive interpolants and maximum entropy principles, in: *Topics in operator theory and interpolation* (Ed. I. Gohberg), OT 29, Birkhäuser Verlag, Basel, 1988, pp. 117-150.

[4] A. Feintuch and B.A. Francis, Distance formulas for operator algebras arising in optimal control problems, in: *Topics in operator theory and interpolation* (Ed. I. Gohberg), OT 29, Birkhäuser Verlag, Basel, 1988, pp. 151-170.

[5] B.A. Francis and J.C. Doyle, Linear control theory with an H_∞ optimality criterion, *SIAM J Control and Optimization* 25 (1987), 815-844.

[6] C. Foias, On an interpolation problem of Dym and Gohberg, *Integral Equations and Operator Theory* 11 (1988), 769-775.

[7] C. Foias and A. Tannenbaum, On the four block problem I, in: *Topics in Operator Theory, Constant Apostel Memorial Issue* (Ed. I. Gohberg), OT 32, Birkhäuser Verlag, Basel, 1988, pp. 93-112.

[8] C. Foias and A. Tannenbaum, On the four block problem II, the singular system, *Integral Equation Operator Theory*, 11 (1988), 726-767.

[9] K. Glover, D.J.N. Limebeer, J.C. Doyle, E.M. Kasenally and M.G. Safonov, A characterization of all solutions to the four block general distance problem, *SIAM J. Control and Optimization* 29 (1991), 283-324.

[10] I. Gohberg, M.A. Kaashoek en H.J. Woerdeman, The band method for positive and contractive extension problems, *J. Operator Theory* 22 (1989), 109-155.

[11] I. Gohberg, M.A. Kaashoek en H.J. Woerdeman, The band method for positive and contractive extension problems: an alternative version and new applications, *Integral Equations Operator Theory* 12 (1989), 343-382.

[12] I. Gohberg, M.A. Kaashoek en H.J. Woerdeman, A maximum entropy principle in the general framework of the band method, *J. Funct. Anal.* 95 (1991), 231-254.

[13] I. Gohberg, M.A. Kaashoek en H.J. Woerdeman, The time variant versions of the Nehari and four block problems, *Springer Lecture Notes*, 1496, pp. 309-323.

J.A. Ball
Department of Mathematics
Virginia Tech
Blacksburg, VA 24061, U.S.A.

I. Gohberg
School of Mathematical Sciences
The Raymond and Beverly Sackler Faculty of Exact Sciences
Tel-Aviv University
Tel-Aviv, Ramat-Aviv 69989, Israel.

M.A. Kaashoek
Faculteit Wiskunde en Informatica
Vrije Universiteit
De Boelelaan 1081a
1081 HV Amsterdam, the Netherlands.

Operator Theory:
Advances and Applications, Vol. 58

THE STATE SPACE METHOD FOR INTEGRO-DIFFERENTIAL EQUATIONS OF WIENER-HOPF TYPE WITH RATIONAL MATRIX SYMBOLS

A. B. Kuijper [1]

This paper concerns systems of integro-differential equations of convolution type on the half-line for which the symbol is a rational matrix function. The equations are studied via certain singular input/output systems. For maximal operators associated with the equations, which act between certain Sobolev spaces on the half-line, explicit conditions for the operators to be invertible and to be Fredholm are derived, as well as explicit formulas for inverses, generalized inverses and the Fredholm characteristics. All conditions and formulas are expressed in terms of the matrices appearing in the singular system corresponding to the equations, and in matrices that are related to the system.

1. INTRODUCTION AND MAIN THEOREMS

During the past ten to fifteen years a new method has been developed to deal with problems in mathematical analysis involving rational matrix-valued functions. This method, which is referred to as the *state space method*, originated in [BGK1] and [BGK2] and is based on the idea of realization, which comes from mathematical systems theory. For a rational $m \times m$ matrix function W which is analytic at infinity, a realization ([Ka]) is a representation in the form

$$W(\lambda) = D + C(\lambda - A)^{-1}B, \tag{1.1}$$

where A is a square matrix of size $n \times n$, say, and B, C and D are matrices of sizes $n \times m$, $m \times n$ and $m \times m$, respectively. The representation (1.1) allows one to reduce analytic problems for W to linear algebra problems involving only the matrices A, B, C and D (see [BGK4] and the references given there).

In this paper the state space method is used to study *systems* of integro-differential equations of the following type:

$$q(i\frac{d}{dt})\phi(t) + \int_0^\infty k(t-s)\phi(s)ds = f(t), \quad 0 < t < \infty. \tag{H}$$

[1] Research supported by the Netherlands organization for scientific research (NWO).

Here $q(\lambda)=\sum_{j=0}^{N} X_j \lambda^j$ is a given polynomial with $m \times m$ matrix coefficients and $k(t)$ is an $m \times m$ matrix function with entries in $L_1(\mathbb{R})$ whose Fourier transform is rational. Thus, the *symbol* of equation (H),

$$W(\lambda)=q(\lambda)+\hat{k}(\lambda),$$

is an arbitrary rational matrix function with no restriction on the behaviour at infinity. Such a function cannot be represented in the form (1.1), but it admits (see [GK1]) a representation

$$W(\lambda)=D+C(\lambda G-A)^{-1}B, \tag{1.2}$$

where A, B, C and D are matrices as before, G is a square matrix of the same size as A and the pencil $\lambda G-A$ is $\mathbb{R}$-regular. The latter means that the pencil $\lambda G-A$ is invertible for all $\lambda \in \mathbb{R}$. In general, the matrix G will not be invertible. On the other hand, because of the freedom one has in the choice of the matrices in (1.2), one may always take $D=I_m$ (see [GK1]).

We use the representation (1.2) with $D=I_m$ to analyse equation (H) in terms of the singular input/output system

$$\begin{cases} iG\rho'(t) &= A\rho(t)+B\phi(t), \\ f(t) &= C\rho(t)+\phi(t). \end{cases} \tag{1.3}$$

Since the feedthrough coefficient is equal to the $m \times m$ identity matrix, we may interchange in (1.3) the role of input and output, and derive the inverse system:

$$\begin{cases} iG\rho'(t) &= A^{\times}\rho(t)+Bf(t), \\ \phi(t) &= C\rho(t)+f(t), \end{cases} \tag{1.4}$$

where $A^{\times}=A-BC$. This approach allows us to obtain explicit descriptions for the Fredholm and invertibility properties of equation (H), and to derive explicit formulas for the solutions in terms of the matrices in (1.2) and the spectral properties of the pencils $\lambda G-A$ and $\lambda G-A^{\times}$ associated with the systems (1.3) and (1.4).

In the following subsections, we shall explain in more detail which equations are considered, and how the state space method is used to obtain solutions.

I.1. The equations. Before turning to the half-line equation (H), we first need some definitions. In the sequel, $(\mathcal{S}^m)'$ will denote the space of $\mathbb{C}^m$-valued tempered distributions ([S]), *i.e.*, in the dual of the space $\mathcal{S}^m$ of $\mathbb{C}^m$-valued rapidly decreasing functions on $\mathbb{R}$. In the setting of $(\mathcal{S}^m)'$ we define an integro-differential operator $\tilde{L}_W:(\mathcal{S}^m)' \to (\mathcal{S}^m)'$ associated with the functions q and k in (H), as follows:

$$\tilde{L}_W\phi := q(i\frac{d}{dt})\phi + k * \phi, \quad \phi \in (\mathcal{S}^m)'. \tag{1.5}$$

Here $\frac{d}{dt}$ and $*$ denote the operator of differentiation and convolution on $(\mathcal{S}^m)'$, respectively, and equality has to be interpreted in distributional sense. More concretely, we have $\tilde{L}_W(\phi)=f$ for a tempered distribution f if and only if

$$\langle \phi, \sum_{j=0}^{N}(-i)^j X_j^T \alpha^{(j)} + \int_{-\infty}^{\infty} k^T(s-t)\alpha(s)ds\rangle = \langle f,\alpha\rangle \tag{1.6}$$

for all test functions $\alpha \in \mathcal{S}^m$. Here $\alpha^{(j)}$ is the jth pointwise derivative of α. For more details on the distributional differentiation $\frac{d}{dt}$ and the convolution operator $*$ on $(\mathcal{S}^m)'$ we refer to [S].

The solutions ϕ of the half-line equation (H) are required to be $\mathbb{C}^m$-valued distributions with *support* in $\overline{\mathbb{R}^+}=[0,\infty)$, *i.e.*,

$$\langle \phi,\alpha\rangle = 0, \quad \alpha\in\mathcal{S}^m, \quad \text{supp } \alpha\subset \mathbb{R}^- = (-\infty,0). \tag{1.7}$$

Then, replacing the integral by a convolution, the left hand side of (H) is a well-defined element of $(\mathcal{S}^m)'$. In the sequel we use the notation $(\mathcal{S}^m)'(\overline{\mathbb{R}^+})$ for the subspace of $(\mathcal{S}^m)'$ of those distributions which satisfy (1.7). Similarly, one defines $(\mathcal{S}^m)'(\overline{\mathbb{R}^-})$ as the subspace of elements in $(\mathcal{S}^m)'$ which have support in $(-\infty,0]$. The equality sign in (H) means that the tempered distribution in the left hand side of (H) coincides on $(0,\infty)$ with the tempered distribution f in the right hand side of (H), in other words, (1.6) holds for all $\alpha\in\mathcal{S}^m$ with supp $\alpha\subset\mathbb{R}^+$.

We shall analyse equation (H) in terms of (possibly unbounded) operators T_W acting from a domain in $H_s^m(\overline{\mathbb{R}^+})$ into $H_{s+r}^m[\mathbb{R}^+]$. Here s and r are integers, $H_s^m(\overline{\mathbb{R}^+})$ is the space of distributions in the Sobolev space $H_s^m(\mathbb{R})$ (cf. [H]) with support in $\overline{\mathbb{R}^+}$, and $H_{s+r}^m[\mathbb{R}^+]$ is the image of $H_s^m(\mathbb{R})$ under the quotient mapping

$$q_+ : (\mathcal{S}^m)' \to (\mathcal{S}^m)'[\mathbb{R}^+] := (\mathcal{S}^m)'/(\mathcal{S}^m)'(\overline{\mathbb{R}^-}), \quad q_+(f)=[f].$$

Now, for prespecified r and s the operator T_W ($H_s^m(\overline{\mathbb{R}^+})\to H_{s+r}^m[\mathbb{R}^+]$) is defined by

$$D(T_W)=\{\phi\in H_s^m(\overline{\mathbb{R}^+}) \mid \tilde{L}_W(\phi)\in H_{s+r}^m(\mathbb{R})+(\mathcal{S}^m)'(\overline{\mathbb{R}^-})\},$$

$$T_W(\phi)=q_+\ \tilde{L}_W(\phi), \quad \phi\in D(T_W).$$

Here the subindex W refers to the symbol $W(\lambda)=q(\lambda)+\hat{k}(\lambda)$ of (H), and $\tilde{L}_W$ is the full line operator defined in (1.5). Observe that T_W is the restriction of the operator $q_+\ \tilde{L}_W : (\mathcal{S}^m)' \to (\mathcal{S}^m)'[\mathbb{R}^+]$ to the largest possible domain in $H_s^m(\overline{\mathbb{R}^+})$ such that Im $T_W\subset H_{s+r}^m[\mathbb{R}^+]$. As a consequence, using the continuity of the operator $\tilde{L}_W$ and the continuity of the embeddings from $H_s^m(\overline{\mathbb{R}^+})$ and $H_{s+r}^m[\mathbb{R}^+]$ into $(\mathcal{S}^m)'$ and $(\mathcal{S}^m)'[\mathbb{R}^+]$, respectively, the operator T_W is closed. The operator T_W will be referred to as the *maximal half-line integro-differential operator with symbol W (acting between $H_s^m(\overline{\mathbb{R}^+})$ and $H_{s+r}^m[\mathbb{R}^+]$).*

Throughout this paper we shall assume the integers $-s$ and $r+s$ to be nonnegative. In this case the right hand side in $T_W(\phi)=[f]$ is a $\mathbb{C}^m$-valued function for which the first $r+s-1$ pointwise derivatives exist as absolutely continuous functions on compact subintervals of $(0,\infty)$, and for which the first $r+s$ pointwise derivatives exist as L_2-functions. On the other hand, the solution ϕ we look for is allowed to be a distribution and need not be a function.

I.2. Singular linear systems. It turns out that the operators $\tilde{L}_W$ and T_W also appear as input/output maps of certain singular linear input/output systems. For the full line operator $\tilde{L}_W$ this connection can be explained by using the Fourier transformation $\mathcal{F}$ in $(\mathcal{S}^m)'$. Applying $\mathcal{F}$ to both sides of $\tilde{L}_W(\phi)=f$ yields $W(\lambda)\mathcal{F}\phi=\mathcal{F}f$. Hence, defining $\rho\in(\mathcal{S}^m)'$ via $\mathcal{F}\rho=(\lambda G-A)^{-1}B\mathcal{F}\phi$, the well-known properties of the Fourier transformation in $(\mathcal{S}^m)'$ (see, *e.g.*, [S]) imply that

$$i\frac{d}{dt}G\rho=A\rho+B\phi,\quad f=C\rho+\phi. \tag{1.8}$$

The corresponding result for the half-line equation (H) reads as follows.

THEOREM 1.1. *Let s and $s+r$ be nonnegative integers. Let the maximal half-line integro-differential operator T_W ($H_s^m(\overline{\mathbb{R}^+})\rightarrow H_{s+r}^m[\mathbb{R}^+]$) have a rational symbol given by*

$$W(\lambda)=I_m+C(\lambda G-A)^{-1}B. \tag{1.9}$$

Assume that $\lambda G-A$ is $\mathbb{R}$-regular, and let $[f]\in H_{s+r}^m[\mathbb{R}^+]$. Then the equation $T_W(\phi)=[f]$ and the singular linear system

$$\begin{cases} i\frac{d}{dt}G\rho &=A\rho+Bu,\\ [y] &=C[\rho]+[u], \end{cases} \tag{Σ}$$

are equivalent in the following sense. If $\phi\in D(T_W)$ is a solution of the equation $T_W(\phi)=[f]$, then the system $(\Sigma]$ with input $u=\phi$ has output $[y]=[f]$. Conversely, let $\phi\in H_s^m(\overline{\mathbb{R}^+})$ and assume that the system $(\Sigma]$ with input $u=\phi$ has output $[y]=[f]$. Then $\phi\in D(T_W)$ and $T_W(\phi)=[f]$.

The statements in the above theorem can be made more precise as follows. Assume that $T_W(\phi)=[f]$ for an element $\phi\in D(T_W)$. Then, the first equation in $(\Sigma]$ with $u=\phi$ has a unique solution ρ in the space $(\mathcal{S}^n)'$, and this solution has the property that $Cq_+(\rho)+q_+(\phi)=[f]$ in the sense of $(\mathcal{S}^m)'[\mathbb{R}^+]$. Conversely, if $\phi\in H_s^m(\overline{\mathbb{R}^+})$ is a distribution for which there exists a $\rho\in(\mathcal{S}^n)'$ satisfying the equations in $(\Sigma]$ with $u=\phi$ and $[y]=[f]$, then $\phi\in D(T_W)$ and $T_W(\phi)=[f]$.

The proof of the above theorem is in Section 6.

Now, let us see how Theorem 1.1 can be used to solve equation (H) or, equivalently, to find the solutions of $T_W(\phi)=[f]$. The first step is to apply the quotient mapping q_+ to the first equation in $(\Sigma]$. This yields

$$i\frac{d}{dt}G[\rho]=A[\rho]+B[\phi], \quad [f]=C[\rho]+[\phi]. \tag{1.10}$$

By interchanging the roles of the input $[\phi]$ and the output $[f]$ in (1.10), one obtains

$$i\frac{d}{dt}G[\rho]=A^{\times}[\rho]+B[f], \quad [\phi]=-C[\rho]+[f]. \tag{1.11}$$

Here $A^{\times}=A-BC$. Next, one solves the first equation in (1.11) for $[\rho]$. At this stage the spectral properties of $\lambda G-A^{\times}$ enter into the analysis. Inserting $[\rho]$ into the second equation of (1.11), one finds $[\phi]$. The final step is to determine representatives $\rho\in(\mathcal{S}^m)'$ of $[\rho]$ and $\phi\in H_s^m(\overline{\mathbb{R}^+})$ of $[\phi]$ satisfying the equation $i\frac{d}{dt}G\rho-A\rho=B\phi$. This final step turns out to be possible only under additional conditions involving the spectral properties of both the pencil $\lambda G-A$ and the pencil $\lambda G-A^{\times}$. These conditions correspond to conditions for the solvability of equation (H).

I.3. Description of results. To state some of our main theorems we need certain spectral projections related to the matrix pencils $\lambda G-A$ and $\lambda G-A^{\times}$ appearing in (1.10) and (1.11). Assume both pencils to be regular. Let V be any subset of $\mathbb{C}$, and let $\Gamma(V)$ be a Cauchy contour around all zeros of $d(\lambda):=\det(\lambda G-A)$ in V, excluding all other zeros of $d(\cdot)$. Put

$$P_V=\frac{1}{2\pi i}\int_{\Gamma(V)}G(\lambda G-A)^{-1}\,d\lambda.$$

The matrices P_V are projections (see [St], [GK1]), and can be viewed as a generalization of the usual Riesz projections. We shall need the projections P_-, P_+ and P_0 which appear by taking V equal to the open upper half plane, the open lower half plane and the real axis, respectively. Also, we need $P_\infty=I-P$, with $P=P_{\mathbb{C}}=P_-+P_++P_0$. Finally, let E and Ω be defined by

$$E=\frac{1}{2\pi i}\int_\Gamma(1-\lambda^{-1})(\lambda G-A)^{-1}\,d\lambda,$$

$$\Omega=\frac{1}{2\pi i}\int_\Gamma(\lambda-\lambda^{-1})G(\lambda G-A)^{-1}\,d\lambda.$$

Here Γ is a Cauchy contour around all zeros of $d(\cdot)$ in the complex plane. The septet

$$(E,\,\Omega,\,P\,;P_-,\,P_+,\,P_0,\,P_\infty) \tag{1.12}$$

will be referred to as the *data for the decomposition of the pencil* $\lambda G-A$. We also need

$$(E^\times, \Omega^\times, P^\times; P_-^\times, P_+^\times, P_0^\times, P_\infty^\times), \tag{1.13}$$

the data for the decomposition of the pencil $\lambda G - A^\times$, which one obtains from the above formulas by replacing A by $A^\times$, where $A^\times = A - BC$. The matrices in (1.12) and (1.13) will be used without further reference.

We now can state the main theorems for the maximal operator related to the half-line equation (H). Recall that the m-dimensional Dirac-distribution $v\delta \in (\mathcal{S}^m)'$, where $v \in \mathbb{C}^m$, is defined by $\langle v.\delta, \alpha\rangle = v^T \alpha(0)$. Furthermore, if g is a $\mathbb{C}^m$-valued function, then g_+ is the function $g_+(t) = g(t)$ for $t \geq 0$ and $g_+(t) = 0$ for $t < 0$; often g_+ is considered as a distribution.

THEOREM 1.2. *Let $s \leq 0$ and $s + r \geq 0$ be integers. Let the maximal half-line integro-differential operator T_W ($H_s^m(\overline{\mathbb{R}^+}) \to H_{s+r}^m[\mathbb{R}^+]$) have a rational symbol W given in the realized form (1.9), with associated decomposition data (1.12)-(1.13). Assume that the pencil $\lambda G - A$ is $\mathbb{R}$-regular. Then the operator T_W is invertible if and only if*

- *(i)* $\lambda G - A^\times$ *is $\mathbb{R}$-regular;*
- *(ii)* $CE^\times(\Omega^\times)^j P_\infty^\times B = 0$ *for $j > r$;*
- *(iii)* $\operatorname{rank} \mathcal{B} = -ms$;
- *(iv)* $\mathbb{C}^n = \operatorname{Im} P_-^\times \oplus \operatorname{Ker} P_- \oplus \operatorname{Im} \mathcal{B}$.

Here $\mathcal{B}$ is the matrix $(B\ \Omega B\ \ldots\ \Omega^{-s-1}B)$. Assume that the above conditions are fulfilled. Choose a left inverse $\mathcal{B}^{(-1)}$ of the matrix $\mathcal{B}$. Then the inverse of T_W is the operator from $H_{s+r}^m[\mathbb{R}^+]$ into $H_s^m(\overline{\mathbb{R}^+})$ defined by

$$\begin{aligned}(T_W)^{-1}(q_+(f)) &= f_+ + \sum_{j=0}^{r+s} CE^\times(i\Omega^\times)^j P_\infty^\times B(f^{(j)})_+ + \\ &+ \sum_{j=r+s+1}^{r} CE^\times(i\Omega^\times)^j P_\infty^\times B(\frac{d}{dt})^{j-r-s}((f^{(r+s)})_+) + \\ &+ (\int_0^\infty k^\times(t-s)f(s)ds - CE^\times e^{-it\Omega^\times}x)_+ + \sum_{j=0}^{-s-1} v_j(i\frac{d}{dt})^j\delta + \\ &+ \sum_{j=0}^{-s-1}\sum_{\nu=j}^{r+s+j-1} CE^\times(i\Omega^\times)^{\nu+1} P_\infty^\times Bf^{(\nu-j)}(0)(\frac{d}{dt})^j\delta,\end{aligned} \tag{1.14}$$

where $f \in H_{s+r}^m(\mathbb{R})$, the vectors x and v_j, $j = 1,2, \cdots, -s-1$, are given by

$$x = -\Pi_1 R_{r,s}(Bf), \quad (v_0^T\ v_1^T\ \ldots\ v_{-s-1}^T)^T = i\mathcal{B}^{(-1)}\Pi_2 R_{r,s}(Bf), \tag{1.15}$$

and $k^\times$ is the function

$$k^\times(t) = \begin{cases} -iCE^\times e^{-it\Omega^\times} P_+^\times B, & t < 0, \\ iCE^\times e^{-it\Omega^\times} P_-^\times B, & t > 0. \end{cases}$$

In the above formulas Π_1 *is the projection of* $\mathbb{C}^n$ *onto* Im $P_-^\times$ *along its natural complement in (iv),* Π_2 *is the projection of* $\mathbb{C}^n$ *onto* Im $\mathcal{B}$ *along its natural complement in (iv), and the mapping* $R_{r,s} : H_{s+r}^n(\mathbb{R}) \to \mathbb{C}^n$ *is given by*

$$R_{r,s}(g) = i\,\Omega^{-s}(\Omega^\times)^{-s}\sum_{j=0}^{r+s-1} P_\infty^\times (i\Omega^\times)^{j+1} g^{(j)}(0) + i\int_0^\infty P_+^\times e^{is\Omega^\times} g(s)ds.$$

THEOREM 1.3. *Let* $s \le 0$ *and* $r+s \ge 0$ *be integers. Let the maximal half-line integro-differential operator* T_W ($H_s^m(\overline{\mathbb{R}^+}) \to H_{s+r}^m[\mathbb{R}^+]$) *have a rational symbol W given in the realized form (1.9), with associated decomposition data (1.12)-(1.13). Assume that the pencil* $\lambda G - A$ *is* $\mathbb{R}$*-regular. Then the operator* T_W *is Fredholm if and only if*

(i) $\lambda G - A^\times$ *is* $\mathbb{R}$*-regular;*

(ii) $CE^\times(\Omega^\times)^j P_\infty^\times B = 0$ *for* $j > r$;

Assume that the above conditions are fulfilled. Then the kernel and image of T_W *are given by*

$$\operatorname{Ker} T_W = \Big\{(CE^\times e^{-it\Omega^\times}x)_+ - \sum_{j=0}^{-s-1} v_j (i\frac{d}{dt})^j \delta \;\Big|\; x \in \operatorname{Im} P_-^\times,\ x + i\sum_{j=0}^{-s-1}\Omega^j B v_j \in \operatorname{Ker} P_-\Big\}, \tag{1.16}$$

$$\dim \operatorname{Ker} T_W = \dim(\operatorname{Im} P_-^\times \cap (\operatorname{Ker} P_- + \operatorname{Im}\mathcal{B})) + \dim(\operatorname{Ker} P_- \cap \operatorname{Im}\mathcal{B}) + \dim \operatorname{Ker}\mathcal{B}. \tag{1.17}$$

$$\operatorname{Im} T_W = \{q_+(f) \;|\; f \in H_{s+r}^m(\mathbb{R}),\ R_{r,s}(Bf) \in \operatorname{Im} P_-^\times + \operatorname{Ker} P_- + \operatorname{Im}\mathcal{B}\}, \tag{1.18}$$

$$\operatorname{codim} \operatorname{Im} T_W = \operatorname{codim}(\operatorname{Im} P_-^\times + \operatorname{Ker} P_- + \operatorname{Im}\mathcal{B}), \tag{1.19}$$

$$\operatorname{ind} T_W = \dim \operatorname{Im} P_-^\times - \dim \operatorname{Im} P_- - ms. \tag{1.20}$$

Here the mapping $R_{r,s} : H_{s+r}^n(\mathbb{R}) \to \mathbb{C}^n$ *and the matrix* $\mathcal{B}$ *are as in Theorem 1.2. Finally, choose a generalized inverse (in the weak sense)* $\mathcal{B}^+$ *of* $\mathcal{B}$*, and let* Λ_1^+ *and* Λ_2^+ *be generalized inverses (in the weak sense) of the operators*

$$\Lambda_1 := (I - P_-^\times)|_{\operatorname{Ker} P_- + \operatorname{Im}\mathcal{B}} : \operatorname{Ker} P_- + \operatorname{Im}\mathcal{B} \to \operatorname{Ker} P_-^\times,$$

$$\Lambda_2 := P_-|_{\operatorname{Im}\mathcal{B}} : \operatorname{Im}\mathcal{B} \to \operatorname{Im} P_-,$$

respectively. Then the operator $T_W^+ : H_{s+r}^m[\mathbb{R}^+] \to H_s^m(\overline{\mathbb{R}^+})$ *given by*

$$
\begin{aligned}
T_W^{+}(q_+(f)) &= f_+ + \sum_{j=0}^{r+s} CE^{\times}(i\Omega^{\times})^j P_{\infty}^{\times} B(f^{(j)})_+ + \\
&+ \sum_{j=r+s+1}^{r} CE^{\times}(i\Omega^{\times})^j P_{\infty}^{\times} B(\frac{d}{dt})^{j-r-s}((f^{(r+s)})_+) + \\
&+ (\int_0^{\infty} k^{\times}(t-s)f(s)ds - CE^{\times}e^{-it\Omega^{\times}}x)_+ + \sum_{j=0}^{-s-1} v_j(i\frac{d}{dt})^j\delta + \\
&+ \sum_{j=0}^{-s-1}\sum_{\nu=j}^{r+s+j-1} CE^{\times}(i\Omega^{\times})^{\nu+1}P_{\infty}^{\times}Bf^{(\nu-j)}(0)(\frac{d}{dt})^j\delta,
\end{aligned}
\tag{1.21}
$$

where $f\in H^m_{s+r}(\mathbb{R})$, *the function* $k^{\times}$ *is as in Theorem 1.2 and the vectors* x *and* v_j, $j=1,2,\ldots,-s-1$, *are given by*

$$x = P_-^{\times}\Lambda_1^{+}(I-P_-^{\times})R_{r,s}(Bf) - P_-^{\times}R_{r,s}(Bf),$$

$$(v_0^T\ v_1^T\ \ldots\ v_{-s-1}^T)^T = i\mathcal{B}^{+}\Lambda_2^{+}P_-\Lambda_1^{+}(I-P_-^{\times})R_{r,s}(Bf),$$

is a generalized inverse of T_W *in the following sense:*

$$T_W T_W^{+} T_W(\phi) = T_W(\phi), \quad \phi\in D(T_W). \tag{1.22}$$

For $s=0$ the conditions (iii) and (iv) in Theorem 1.2 reduce to the matching condition

$$\mathbb{C}^n = \operatorname{Im} P_-^{\times} \oplus \operatorname{Ker} P_-.$$

Also, the formulas (1.14)-(1.21) simplify considerably in this case, because all summations with upper bound $-s-1$ vanish and the matrix $\mathcal{B}$ has size 0×0. When $s<0$, the right hand side of formula (1.14) does not depend on the specific choice of the left inverse $\mathcal{B}^{(-1)}$ of $\mathcal{B}$. Indeed, for an injective operator X all left inverses coincide on Im X. Thus, formula (1.14) is fully determined by the realization (1.9) of W and the right hand side $[f]$ of equation (H). Furthermore, the conditions (i) and (ii) may be replaced by two equivalent conditions (i') and (ii'):

(i') det $W(\lambda)\neq 0$ *for* $\lambda\in\mathbb{R}$;

(ii') $W(\lambda)^{-1}\lambda^{-r}$ *is bounded.*

Note that condition (ii') is automatically fulfilled for r sufficiently large. It seems to be an open question whether conditions (i') and (ii') are also necessary and sufficient for the (possibly unbounded) integro-differential operator T_W to be Fredholm if the kernel function k is a general $m\times m$ matrix-valued function with components in $L_1(\mathbb{R})$. Finally, we point out that the Fredholm indices Ker T_W, codim Im T_W and ind T_W depend only on the parameter s.

The results obtained in this paper have various predecessors. By taking $q(\lambda)\equiv I_m$, the equations studied here reduce to Wiener-Hopf integral equations of the second kind with a rational symbol. Equations of the latter type have been explicitly solved in [BGK1] by using the realization (1.1) (see also [GGK], Chapter XIII and Section XVIII.5). In fact, the results in this paper may be viewed as generalizations of the earlier results in [BGK1]. The class of equations (H) also covers systems of ordinary differential equations with constant coefficients, corresponding to both monic and nonmonic matrix polynomials, for which explicit formulas for solutions in terms of standard triples have been obtained in [GKLR] and [GLR]. In [GKLR] and [GLR] the right hand side is taken sufficiently smooth and the solutions are functions not necessarily restricted to a prescribed function space. First kind Wiener-Hopf equations with rational matrix symbols, *i.e.*, the case when $q(\lambda) \equiv 0$, were explicitly solved earlier in [Ro] by using the realization (1.1) with $D = 0$. A standard assumption in [Ro] is that $W(\lambda)\lambda^r$ is proper, which allows one to remain within the framework of bounded operators. In the present paper, for the first time all the equations mentioned above are treated in terms of the state space method from one point of view.

Realizations of type (1.9) have been used earlier in [GK1] and [GK2] to solve Toeplitz equations and singular integral equations with rational matrix symbols. In both cases the symbol is regular on the full Cauchy contour associated with the equations. In our case an essential difficulty is that on the extended real line the symbol has a singularity at infinity.

An inversion theory of *scalar* integro-differential equations of convolution type, not necessarily with a rational symbol, is developed in [Ta], [VG1] and [VG2] (see also [VG3]-[VG5]). In these papers the main emphasis is on generalizations of the factorization method developed in the earlier work of [Kr] for second kind Wiener-Hopf integral equations (see also [GKr] for systems of second kind equations). More precisely, under the assumption that the symbol of the equation admits a certain factorization, Talenti [Ta] shows that the natural operator corresponding to the equation acting between well-chosen Sobolev spaces is Fredholm. Furthermore, Talenti describes the Fredholm characteristics of the operator in terms of the factorization. Under the assumption of a slightly more general factorization of the symbol, it is shown in [VG1]-[VG2] (see also [VG3]-[VG5]) that one can associate to the equation an in a certain sense invertible operator acting between well-chosen scales of Sobolev spaces. The factorizations in [Ta] and [VG1]-[VG5] always exist for rational functions without poles and zeros on the real line. Another approach for first kind equations with positive definite symbols, which also works for the matrix case, appears in [R1] (see also [R2]). All the above mentioned references present a method of obtaining solutions, *e.g.* via factorization, but, in general, they do not give explicit formulas.

I.4. Descriptions of contents. This paper contains a large part of the author's thesis [Ku]. Not counting the present introduction, the paper consists of seven sections. Section 2 contains preliminaries on the spectral properties of matrix pencils. Also, the decomposition data for a pencil are introduced in Section 2. Sections 3 and 4 deal with singular differential equations, *i.e.*, differential equations of the types appearing in the first parts of (1.8) and (1.10). The general solutions are described in terms of the data for the decomposition of the pencil associated with the equation. Section 5 concerns preliminaries on realizations. A number of auxiliary results that will play an important role later are presented here. Section 6 contains the proof of Theorem 1.1. The proofs of Theorems 1.2 and 1.3 is in Section 7. Finally, Section 8 contains an example illustrating the earlier results.

2. PRELIMINARIES ON MATRIX PENCILS

Given two square matrices G and A of size $n \times n$ say, the expression $\lambda G - A$, where λ is a complex parameter, is called a *(linear matrix) pencil*. This section concerns the spectral properties of $\lambda G - A$. We begin with some definitions. A pencil is said to be *regular* if its determinant is not identically equal to zero. For a nonempty subset Δ of the Riemann sphere $\mathbb{C}_\infty = \mathbb{C} \cup \{\infty\}$, a pencil $\lambda G - A$ is Δ*-regular* if $\lambda G - A$ (or just G if $\lambda = \infty$) is invertible for each λ in Δ. Clearly, Δ- regularity of a pencil implies regularity. Whenever necessary an $n \times n$ matrix is identified with the operator on $\mathbb{C}^n$ of which the matrix relative to the standard basis in $\mathbb{C}^n$ is the given one.

Three matrices that are closely related to linear pencils will occur frequently in this paper. Let $\lambda G - A$ be a regular matrix pencil. We define the *left separating projection P*, the *right equivalence matrix E* and the *associate matrix* Ω of the pencil $\lambda G - A$ (corresponding to ∞) by

$$P = \frac{1}{2\pi i} \int_\Gamma G(\zeta G - A)^{-1} \, d\zeta,$$

$$E = \frac{1}{2\pi i} \int_\Gamma (1 - \zeta^{-1})(\zeta G - A)^{-1} \, d\zeta,$$

$$\Omega = \frac{1}{2\pi i} \int_\Gamma (\zeta - \zeta^{-1}) G(\zeta G - A)^{-1} \, d\zeta.$$

Here Γ is a Cauchy contour such that all the zeros of det $(\lambda G - A)$ are in the interior domain Δ_+ of Γ. Also, we assume that 0 is in the interior domain of Γ. Apart from this, the particular choice of the contour Γ is irrelevant.

The above terminology has been taken from [GK1]. In [GK1], Theorem 2.1, it is shown that the matrix E is invertible and that the matrix P is idempotent. Furthermore, in [GK1] the following identities are proved (cf. [GK1], Lemma 2.2 and Theorem 2.1):

$$AE = \Omega P + (I - P), \quad GE = P + \Omega(I - P). \tag{2.1}$$

Note that these identities imply that $(\lambda G - A)E$ leaves the subspaces Ker P and Im P invariant. In fact, the matrices P, E and Ω induce a decomposition of the pencil $\lambda G - A$ of the following form:

$$(\lambda G - A)E = \begin{bmatrix} \lambda\Omega_1 - I_1 & 0 \\ 0 & \lambda I_2 - \Omega_2 \end{bmatrix} : \mathrm{Ker}\, P \oplus \mathrm{Im}\, P \to \mathrm{Ker}\, P \oplus \mathrm{Im}\, P, \tag{2.2}$$

$$\Omega = \begin{bmatrix} \Omega_1 & 0 \\ 0 & \Omega_2 \end{bmatrix} : \mathrm{Ker}\, P \oplus \mathrm{Im}\, P \to \mathrm{Ker}\, P \oplus \mathrm{Im}\, P, \tag{2.3}$$

where $\lambda\Omega_1 - I_1$ is a $(\Delta_+ \cup \Gamma)$-regular pencil and $\lambda - \Omega_2$ is a $(\mathbb{C}_\infty \setminus \Delta_+)$-regular pencil. Moreover, P and E are the unique operators such that the partitions in (2.2) and (2.3) hold true. These properties justify the terminology for the matrices P, E and Ω. Note that, in contrast to [GK1], we define the matrices P, E and Ω only for a Cauchy contour Γ around all the zeros of det $(\lambda G - A)$.

The next two lemmas provide additional information about the matrices P, E and Ω. Recall that the *index of nilpotency* of a nilpotent matrix N is the smallest integer δ such that $N^\delta = 0$. For a rational $m \times m$ matrix function W with determinant not identically equal to zero, the *zeros* of W are the points in $\mathbb{C}_\infty$ where the rational matrix function $W^{-1}(\lambda) := W(\lambda)^{-1}$ has its poles. For a regular pencil $\lambda G - A$ this implies that the zeros inside the complex plane are precisely the points where det $(\lambda G - A) = 0$. The *order* of the pole at infinity of a rational matrix function W is the smallest number $\nu \geq 0$ such that $\lambda^{-\nu} W(\lambda)$ is analytic at infinity. Finally, the spectrum of an operator (a matrix) X will be denoted by $\sigma(X)$.

LEMMA 2.1. *Let $\lambda G - A$ be a regular $n \times n$ matrix pencil. Let P and Ω be the left separating projection and the associate matrix of the pencil $\lambda G - A$, respectively. Then the matrix $\Omega(I - P)$ is nilpotent and the index of nilpotency of $\Omega(I - P)$ is equal to the order of the pole at infinity of the function $(\lambda G - A)^{-1}$ plus one.*

PROOF. The proof is based on the decomposition of the pencil $\lambda G - A$ in (2.2) and (2.3). Recall that the pencil $\lambda\Omega_1 - I_1$ is invertible for all $\lambda \in \Gamma \cup \Delta_+$. Since all the zeros of $\lambda G - A$ in $\mathbb{C}$ are in Δ_+, formula (2.2) implies that the pencil $\lambda\Omega_1 - I_1$ is also regular on $\mathbb{C} \setminus \Delta_+$. Thus, $\lambda\Omega_1 - I_1$ is invertible for all λ in $\mathbb{C}$, and hence Ω_1 is a nilpotent matrix. Moreover, taking inverses on both sides of the equation (2.2), one immediately sees that the index of nilpotency of the matrix Ω_1 equals the order of the pole at infinity of the function $(\lambda G - A)^{-1}$ plus one. Using the decomposition in (2.3), the same remarks hold for the matrix $\Omega(I - P)$, and the lemma is proved. □

LEMMA 2.2. *Let $\lambda G - A$ be a regular $n \times n$ matrix pencil. For an arbitrary subset V of $\mathbb{C}$, define the matrix P_V by*

$$P_V = \frac{1}{2\pi i} \int_{\Gamma(V)} G(\zeta G - A)^{-1} d\zeta,$$

where $\Gamma(V)$ is a Cauchy contour around all zeros of $\lambda G - A$ in V such that all other zeros of $\lambda G - A$ are outside $\Gamma(V)$. Let P and Ω be the left separating projection and the associate matrix of the pencil $\lambda G - A$, respectively. Then, P_V is an idempotent matrix satisfying $PP_V = P_V P = P_V$. Furthermore, $\Omega P_V = P_V \Omega$, and the spectrum of $\Omega|_{\mathrm{Im}\, P_V}$ coincides with the set of zeros of the pencil $\lambda G - A$ inside V.

PROOF. The well known contour integration arguments together with the generalized resolvent identity for pencils,

$$(\lambda G - A)^{-1} - (\mu G - A)^{-1} = (\mu - \lambda)(\lambda G - A)^{-1} G (\mu G - A)^{-1},$$

show that $P_V^2 = P_V$ and $PP_V = P_V P = P_V$. Next, use equation (2.2) to compute P_V with respect to the decomposition $\mathbb{C}^n = \mathrm{Ker}\, P \oplus \mathrm{Im}\, P$. This yields

$$P_V = \frac{1}{2\pi i} \int_{\Gamma(V)} GE\left[(\zeta G - A)E\right]^{-1} d\zeta =$$

$$\frac{1}{2\pi i} \int_{\Gamma(V)} \begin{bmatrix} (\zeta\Omega_1 - I_1)^{-1}\Omega_1 & 0 \\ 0 & (\zeta I_2 - \Omega_2)^{-1} \end{bmatrix} d\zeta = \begin{bmatrix} 0 & 0 \\ 0 & P(\Omega_2; V) \end{bmatrix},$$

where $P(\Omega_2; V)$ is the spectral projection of the matrix Ω_2 corresponding to the part of $\sigma(\Omega_2)$ inside V. To justify the last step, note that the (1,1) block element in the integrand is analytic, which is a consequence of the nilpotency of the matrix Ω_1. Now, the above calculations, together with (2.3) imply that $\Omega P_V = P_V \Omega$. Moreover, it follows that $\sigma(\Omega|_{\mathrm{Im}\, P_V})$ consists of the eigenvalues of the matrix Ω_2 inside V. Formula (2.2) now proves the lemma. □

We shall refer to the matrix P_V defined above as the *left spectral projection corresponding to* V of the pencil $\lambda G - A$. Note that it is not a priori assumed that $0 \in V$. In case $V = \mathbb{C}$, then $P_V = P$, the left separating projection of $\lambda G - A$. If $V = \mathbb{C}_+$ ($\mathbb{C}_-$), i.e. V is the open upper (lower) half plane, we use the notation P_+ (P_-) for the corresponding left spectral projection. If $V = \mathbb{R}$ we write $P_V = P_0$. Finally, we introduce the *left spectral projection corresponding to* ∞, $P_\infty = I - P$. We shall henceforth refer to the ordered set of seven operators

$$(E, \Omega, P; P_-, P_+, P_0, P_\infty) \tag{2.4}$$

as the *data for the decomposition of the pencil* $\lambda G - A$.

When the matrix G is the $n \times n$ identity matrix, then the zeros of the pencil $\lambda G - A$ are precisely the eigenvalues of the matrix A, and we have $P = I_n$, $E = I_n$ and $\Omega = A$. In this case Lemma 2.1 is a lemma about empty matrices and the second lemma is nothing else but a corollary of the classical Riesz theory. The extension of this theory to regular (matrix and operator) pencils is due to in [St]. See also [GGK], Section IV.1.

3. SINGULAR DIFFERENTIAL EQUATIONS ON THE FULL-LINE

Let $\lambda G - A$ be a given $n \times n$ regular matrix pencil. In this section we consider the equation

$$i\frac{d}{dt}G\rho - A\rho = \phi \tag{3.1}$$

in the context of the space of tempered distributions $(\mathcal{S}^n)'$, the dual of the space $\mathcal{S}^n$. In particular, given an element ϕ in $(\mathcal{S}^n)'$, we look for all solutions of equation (3.1) in the same space. An element $\rho \in (\mathcal{S}^n)'$ is called a *solution* of equation (3.1) if

$$-i\langle \rho, G^T \alpha' \rangle = \langle \rho, A^T \alpha \rangle + \langle \phi, \alpha \rangle, \qquad \alpha \in \mathcal{S}^n,$$

where α' indicates the classical derivative of the function α.

THEOREM 3.1. *Let $\lambda G - A$ be an $n \times n$ regular matrix pencil, and let $(E, \Omega, P; P_-, P_+, P_0, P_\infty)$ be the data for the decomposition of $\lambda G - A$. Let ϕ be an arbitrary element of $(\mathcal{S}^n)'$. Then the general solution of equation (3.1) in the space $(\mathcal{S}^n)'$ is given by*

$$\rho = l * \phi + p(i\frac{d}{dt})\phi + L(\phi) + Ee^{-it\Omega}x, \tag{3.2}$$

where x is an arbitrary vector in Im P_0, *the functions l and p are given by*

$$l(t) = \begin{cases} iEe^{-it\Omega}P_+, & t < 0, \\ -iEe^{-it\Omega}P_-, & t > 0, \end{cases}$$

$$p(\lambda) = -\sum_{j=0}^{N} E\Omega^j P_\infty \lambda^j,$$

$L(\phi) = Ee^{-i\Omega t}P_0 F \in (\mathcal{S}^n)'$ *with F a primitive in $(\mathcal{S}^n)'$ of $-ie^{i\Omega t}P_0\phi$, and N is the order of the pole at infinity of $(\lambda G - A)^{-1}$.*

PROOF. The proof is in five steps.

Step (i). In the first step we show that the right hand side of (3.2) is a well-defined tempered distribution and a solution of (3.1). From Lemma 2.2 we know that $l(t)$ is an exponentially decreasing function for $t \to \pm\infty$. Hence by standard results from the theory of distributions, the convolution in (3.2) is well defined in $(\mathcal{S}^n)'$. Furthermore, Lemma 2.2 shows that $e^{i\Omega t}P_0$ is a $\mathcal{C}^\infty$ matrix function bounded in norm by a polynomial. Thus, $f = ie^{i\Omega t}P_0\phi$ is a tempered distribution, and we may choose a primitive F of f in $(\mathcal{S}^n)'$ (cf. [S], Theorem II,I and Theorem VII,VI). Next, choose a vector x from the subspace Im P_0, and define the distribution ρ by the right hand side of (3.2). Then ρ is well-defined. To prove that ρ satisfies (3.1), first recall the following identities from Section 2:

$$P_\infty = I - P, \quad P_\nu = PP_\nu = P_\nu P, \quad \nu \in \{+,0,-\}.$$

Put $\rho_p = E^{-1} p(i\frac{d}{dt})\phi$, and $\rho_l = E^{-1}(l * \phi)$. Then, as $\Omega^{N+1}(I-P)=0$ by Lemma 2.1, we have

$$i\frac{d}{dt}\Omega\rho_p = -\sum_{j=0}^{N}(i\Omega)^{j+1}P_\infty(\frac{d}{dt})^{j+1}\phi = -\sum_{j=1}^{N+1}(i\Omega)^j P_\infty(\frac{d}{dt})^j\phi =$$

$$= -\sum_{j=0}^{N}(i\Omega)^j P_\infty(\frac{d}{dt})^j\phi + P_\infty\phi = \rho_p + P_\infty\phi.$$

Furthermore, standard properties of the convolution product and the differentiation in the space $(\mathcal{S}^n)'$ show that

$$i\frac{d}{dt}\rho_l = i\frac{d}{dt}E^{-1}(l * \phi) = i(\frac{d}{dt}E^{-1}l) * \phi =$$

$$= \Omega(E^{-1}l) * \phi + (P_- + P_+)\phi = \Omega\rho_l + (P_- + P_+)\phi.$$

Using formula (2.1), a combination of the above two formulas yields

$$i\frac{d}{dt}GE(\rho_p + \rho_l) - AE(\rho_p + \rho_l) =$$

$$= i\frac{d}{dt}\Omega\rho_p - \rho_p + i\frac{d}{dt}\rho_l - \Omega\rho_l = (P_\infty + P_+ + P_-)\phi = (I - P_0)\phi.$$

The next step is to consider the third term in (3.2). Compute that

$$i\frac{d}{dt}GL(\phi) - AL(\phi) = i\frac{d}{dt}GEe^{-i\Omega t}P_0F - AEe^{-i\Omega t}P_0F =$$

$$= iGE(-i\Omega)e^{-i\Omega t}P_0F - iGEe^{-i\Omega t}ie^{i\Omega t}P_0\phi - AEe^{-i\Omega t}P_0F = P_0\phi.$$

Finally, a straightfoward calculation shows that the fourth term in (3.2) does not contribute to the left hand side in (3.1) at all, as

$$i\frac{d}{dt}GEe^{-it\Omega}x - AEe^{-it\Omega}x = i\frac{d}{dt}e^{-it\Omega}x - \Omega e^{-it\Omega}x = 0.$$

Here we used the results of Section 2, again. A combination of the above arguments now prove that ρ is a solution of (3.1), indeed.

Step (ii). Fix a primitive F of $-ie^{it\Omega}P_0\phi$ and a vector x_0 in Im P_0, and put ρ_0 the distribution as in (3.2) with $x = x_0$. Suppose that $\tilde{\rho}_0 \in (\mathcal{S}^n)'$ is another solution to equation (3.1). Then $\rho_0 - \tilde{\rho}_0$ is a solution of the homogeneous equation

$$i\frac{d}{dt}G\rho = A\rho. \tag{3.3}$$

Thus, to finish the proof of Theorem 3.1, it suffices to show that any solution of (3.3) in the space $(\mathcal{S}^n)'$ is of the form

$$\rho(t) = Ee^{-i\Omega t}Px, \qquad -\infty < t < \infty, \tag{3.4}$$

where x is a vector in Im P_0. As a preparation to solving equation (3.3), we first study two special cases of equation (3.3) in the space $(\mathcal{D}^n)'$ of n-dimensional distributions (see [S], [H]), a space which contains the space $(\mathcal{S}^n)'$.

Step (iii). In this step we consider equations of type

$$i\frac{d}{dt}X\psi = \psi, \tag{3.5}$$

where X is a nilpotent matrix of size $k \times k$, say, in the space $(\mathcal{D}^k)'$. We call $\psi \in (\mathcal{D}^k)'$ a solution of (3.5) if

$$-i\langle\psi, X^T\alpha'\rangle = \langle\psi,\alpha\rangle, \qquad \alpha \in \mathcal{D}^k,$$

where $\mathcal{D}^k$ is the space of $\mathbb{C}^k$-valued infinitely differentiable functions with compact support (see [S]), and α' denotes the pointwise classical derivative of the function α. Clearly, $\psi=0$ is a solution to equation (3.5). We claim that it is, in fact, the only solution of (3.5). Indeed, let $\tilde{N}$ be the order of nilpotency of the matrix X. A premultiplication of the equality in (3.5) with $X^{\tilde{N}-1}$ yields $X^{\tilde{N}-1}\psi=0$. Thus $X^{\tilde{N}-1}(\frac{d}{dt}\psi)=0$. But then, a premultiplication of identity (3.5) with $X^{\tilde{N}-2}$ shows that $X^{\tilde{N}-2}\psi=0$. Continuing like this, we eventually have that $\psi=X^0\psi=0$, which proves the claim.

Step (iv). Now, consider an equation of the type

$$i\frac{d}{dt}\psi = Y\psi, \tag{3.6}$$

where Y is a $k \times k$ matrix, in the setting of $(\mathcal{D}^k)'$. We call $\psi \in (\mathcal{D}^k)'$ a solution of (3.6) if

$$-i\langle\psi,\alpha'\rangle = \langle\psi, Y^T\alpha\rangle, \qquad \alpha \in \mathcal{D}^k.$$

The general solution to the above equation is well known (see, *e.g.*, [S], Theorem V,IX) and consists of the solutions of equation (3.6) interpreted in the usual sense,

$$\psi(t) = e^{-itY}w, \qquad -\infty < t < \infty.$$

Step (v). Let ρ be a solution in the space $(\mathcal{S}^n)'$ of equation (3.3). Put $\rho_1 = PE^{-1}\rho$ and

$\rho_2 = (I-P)E^{-1}\rho$. Then $\rho = E\rho_1 + E\rho_2$, and a premultiplication of the identity in (3.3) with P, respectively $I-P$ shows that

$$i\frac{d}{dt}\rho_1 = \Omega P\rho_1, \quad i\frac{d}{dt}\Omega(I-P)\rho_2 = \rho_2. \tag{3.7}$$

Here we used the formulas in (3.1), the fact that $\Omega P = P\Omega$ (cf. Lemma 2.2), and the trivial observation that $P\rho_1 = \rho_1$, $P\rho_2 = 0$. Now, recall that $(\mathcal{S}^n)' \subset (\mathcal{D}^n)'$. So, ρ_1 and ρ_2 are distributions in $(\mathcal{D}^n)'$ which satisfy the equations in (3.7) in $(\mathcal{D}^n)'$-sense. But then, as $\Omega(I-P)$ is a nilpotent matrix by Lemma 2.1, Step (iii) of the proof shows that $\rho_2 = 0$. Furthermore, it follows from Step (iv) that the general solution to the first equation in (3.7) is given by

$$\rho_1(t) = P\rho_1(t) = e^{-i\Omega P t}Px = e^{-i\Omega t}Px, \qquad -\infty < t < \infty,$$

with x an arbitrary vector in $\mathbb{C}^n$. Thus, $\rho = E\rho_1 + E\rho_2 = E\rho_1$ is of the form

$$\rho(t) = Ee^{-it\Omega}Px, \tag{3.8}$$

for some vector x in $\mathbb{C}^n$. Put $P_e = P_+ + P_-$. Then $P = P_0 + P_e$, and we may write $\rho = \rho_0 + \rho_e$. Here ρ_ν is the distribution in $(\mathcal{D}^n)'$ given by the right hand side of (3.8) with P_ν instead of P, $\nu \in \{0, e\}$. Now, recall from Lemma 2.2 that

$$\sigma(\Omega|_{\operatorname{Im} P_0}) \subset \mathbb{R}, \qquad \sigma(\Omega|_{\operatorname{Im} P_e}) \cap \mathbb{R} = \varnothing.$$

As a consequence, ρ_0 is a continuous function bounded in norm by a polynomial. In other words, using a well known characterization of tempered distributions (see [S], Theorem VII,VI), ρ_0 is an element of $(\mathcal{S}^n)'$. But then, $\rho_e \in (\mathcal{S}^n)'$ as well, and by the same characterization of $(\mathcal{S}^n)'$ we must have that $P_e x = 0$. We conclude that $\rho = \rho_0$ is of the form indicated in the theorem. This finishes step (v). □

In case the distribution ϕ is a tempered function, *i.e.*, a function bounded in norm by a polynomial, one possible choice for $L(\phi)$ is the function:

$$L(\phi) = -iE\int_0^t e^{-i(t-s)\Omega}P_0\phi(s)ds.$$

Hence, in this case the general solution to equation (3.1) is given by

$$\rho = l * \phi + p(i\frac{d}{dt})\phi - iE\int_0^t e^{-i(t-s)\Omega}P_0\phi(s)ds + Ee^{-it\Omega}x,$$

where the functions $p(\lambda)$ and $k(t)$ and the vector x are as in Theorem 3.1.

We conclude this section with the specification of Theorem 3.1 for $\mathbb{R}$-regular pencils.

COROLLARY 3.2. *Let $\lambda G - A$ be an $n \times n$ regular matrix pencil. Then equation (3.1)* *has a unique solution for every right hand side* $\phi \in (\mathcal{S}^n)'$ *if and only if* $\lambda G - A$ *is* $\mathbb{R}$*-regular, and in this case the solution is given by*

$$\rho = l * \phi + p(i\frac{d}{dt})\phi, \tag{3.9}$$

where the functions p *and* l *are as in Theorem 3.1.*

PROOF. From the proof of Theorem 3.1 it follows that the general solution of the homogeneous equation

$$i\frac{d}{dt}G\rho - A\rho = 0$$

in $(\mathcal{S}^n)'$ is given by

$$\rho(t) = Ee^{-it\Omega}x, \qquad -\infty < t < \infty,$$

where x runs over all vectors in Im P_0, the image of the left spectral projection of $\lambda G - A$ corresponding to the real axis. Thus, equation (3.2) has a unique solution in $(\mathcal{S}^n)'$ for every right hand side $\phi \in (\mathcal{S}^n)'$ if and only if $P_0 = 0$, or in other words if $\lambda G - A$ is $\mathbb{R}$-regular. In this case, the formula (3.9) for the solution immediately follows from (3.2). □

The solution (3.9) can also be obtained via the frequency domain by applying the Fourier transform to (3.1).

4. SINGULAR DIFFERENTIAL EQUATIONS ON THE HALF-LINE

Let $\lambda G - A$ be a regular $n \times n$ matrix pencil. In this section we consider the equation

$$i\frac{d}{dt}G[\rho] = A[\rho] + [\phi] \tag{4.1}$$

in the context of the space $(\mathcal{S}^n)'[\mathbb{R}^+]$, the quotient space $(\mathcal{S}^n)'/(\mathcal{S}^n)'(\overline{\mathbb{R}^-})$ of tempered distributions in $(0,\infty)$. The corresponding quotient mapping we denote by q_+, and equivalence classes in $(\mathcal{S}^n)'[\mathbb{R}^+]$ are denoted by square brackets, like in $[\rho]$. Given an element $[\phi]$ in $(\mathcal{S}^n)'[\mathbb{R}^+]$, an element $[\rho] \in (\mathcal{S}^n)'[\mathbb{R}^+]$ is called a *solution* of equation (4.1) if

$$-i\langle[\rho], G^T\alpha'\rangle = \langle[\rho], A^T\alpha\rangle + \langle[\phi], \alpha\rangle, \quad \alpha \in \mathcal{S}^n, \ \text{supp}(\alpha) \subset (0,\infty),$$

where α' indicates the classical derivative of $\alpha \in \mathcal{S}^n$.

THEOREM 4.1. *Let* $\lambda G - A$ *be an* $n \times n$ *regular matrix pencil, and let* $(E, \Omega, P; P_-, P_+, P_0, P_\infty)$ *be the data for the decomposition of* $\lambda G - A$. *Let* $[\phi]$ *be an arbitrary element of* $(\mathcal{S}^n)'[\mathbb{R}^+]$. *Then the general solution of equation (4.1) in the space* $(\mathcal{S}^n)'[\mathbb{R}^+]$ *is given by*

$$[\rho]=q_+\big(l*\psi+p(i\frac{d}{dt})\psi+L(\psi)\big)+\big[Ee^{-it\Omega}x\big], \tag{4.2}$$

where x is an arbitrary vector in $\mathrm{Im}\, P_- + \mathrm{Im}\, P_0$, *the element* $\psi \in (\mathscr{S}^n)'$ *is any distribution such that* $q_+(\psi)=[\phi]$, *the functions l and p are given by*

$$l(t)=\begin{cases} iEe^{-it\Omega}P_+, & t<0,\\ -iEe^{-it\Omega}P_-, & t>0,\end{cases}$$

$$p(\lambda)=-\sum_{j=0}^{N} E\Omega^j P_\infty \lambda^j,$$

$L(\psi)=Ee^{-i\Omega t}P_0F \in (\mathscr{S}^n)'$ *with F a primitive in* $(\mathscr{S}^n)'$ *of* $-ie^{i\Omega t}P_0\psi$, *and N is the order of the pole at infinity of* $(\lambda G-A)^{-1}$.

PROOF. From Theorem 3.1 we know that

$$\rho_1=l*\psi+p(i\frac{d}{dt})\psi+L(\psi)$$

is a solution of the equation $i\frac{d}{dt}G\rho=A\rho+\psi$ in the space $(\mathscr{S}^n)'$. Hence, $q_+(\rho_1)$ is a solution of equation (4.1). Also, $[Ee^{-it\Omega}x]$ is an element of $(\mathscr{S}^n)'[\mathbb{R}^+]$ and satisfies

$$i\frac{d}{dt}G\big[Ee^{-it\Omega}x\big]-A\big[Ee^{-it\Omega}x\big]=\big[i\frac{d}{dt}e^{-it\Omega}x\big]-\big[\Omega e^{-it\Omega}x\big]=[0].$$

Thus, every $[\rho]$ of the form (4.2) is an element of $(\mathscr{S}^n)'[\mathbb{R}^+]$ and a solution of (4.1).

To prove the converse, it suffices to show that every solution of the homogeneous equation

$$i\frac{d}{dt}G[\rho]=A[\rho],\quad [\rho]\in(\mathscr{S}^n)'[\mathbb{R}^+], \tag{4.3}$$

is of the form $[Ee^{-it\Omega}x]$, with x as in the theorem. Let $[\rho]$ be a solution of (4.3). Put $[\rho]=E[\rho_1]+E[\rho_2]$, with $[\rho_1]=PE^{-1}[\rho]$ and $[\rho_2]=P_\infty E^{-1}[\rho]$. By straightforward verification one shows that $[\rho_1]$ and $[\rho_2]$ satisfy the following equations:

$$i\frac{d}{dt}[\rho_1]=\Omega P[\rho_1],\quad i\frac{d}{dt}\Omega P_\infty[\rho_2]=[\rho_2]. \tag{4.4}$$

The key remark to finish the proof now is to repeat the arguments in steps (iii) and (iv) in the proof of Theorem 3.1. Firstly, note that $[\rho_1]$ and $[\rho_2]$, being tempered distributions in $(0,\infty)$, may be viewed as elements of $\mathscr{D}^n(\mathbb{R}^+)'$, the space of distributions in $(0,\infty)$ (see [H], where the space $\mathscr{D}^n(\mathbb{R}^+)'$ is denoted by $(\mathscr{D}^n)'(\mathbb{R}^+)$). Moreover, the equations in (4.4) are satisfied in

$\mathscr{D}^n(\mathbb{R}^+)'$-sense. Since this space has exactly the same structure as $(\mathscr{D}^n)'$ itself, the arguments in the proof of Theorem 3.1, Step (iii) and (iv), still hold. Therefore $[\rho_2]=0$. Furthermore, $[\rho_1]$ will be of the form

$$[\rho_1]=P[\rho_1]=[e^{-it\Omega}\tilde{x}],$$

where $\tilde{x}$ is a vector in Im P. However, $[\rho_1]\in(\mathscr{S}^n)'[\mathbb{R}^+]$, which implies that $x=\tilde{x}\in$ Im P_-+Im P_0. Thus $[\rho]=E[\rho_1]$ is of the desired form. □

To conclude this section, we specify the above result for the case when ϕ is in $L_2^n(\overline{\mathbb{R}^+})$, the space of $\mathbb{C}^n$-valued L_2-functions with support in $[0,\infty)$. For such a function, we indicate by $[\phi]$ the corresponding element in $(\mathscr{S}^n)'[\mathbb{R}^+]$.

COROLLARY 4.2. *Let $\lambda G-A$ be an regular $n\times n$ matrix pencil. Let ϕ be a given function in $L_2^n(\overline{\mathbb{R}^+})$. Then the general solution of equation* (4.1) *in $(\mathscr{S}^n)'[\mathbb{R}^+]$ is given by*

$$[\rho]=\Big[\int_0^\infty l(t-s)\phi(s)ds\Big]+p(i\frac{d}{dt})[\phi]+$$

$$-\Big[iE\int_0^t e^{-i(t-s)\Omega}P_0\phi(s)ds\Big]+\Big[Ee^{-it\Omega}x\Big],\quad x\in \text{Im } P_-+\text{Im } P_0.$$

Here $l, p, P_-, P_0, E,$ and Ω are as in Theorem 4.1.

5. PRELIMINARIES ON REALIZATIONS

This section deals with a special representation for rational matrix functions. A triple $\Theta=(\lambda G-A,B,C)$ is called a *realization* of a given rational $m\times m$ matrix function W if

$$W(\lambda)=I+C(\lambda G-A)^{-1}B. \tag{5.1}$$

Here $\lambda G-A$ is a regular matrix pencil of order n , say, and B and C are matrices of sizes $n\times m$ and $m\times n$, respectively. In [GK1] an explicit procedure to build a realization for any rational $m\times m$ matrix function is described. This procedure is based on the Laurent series development of a function $W(\lambda)$ at its poles, and yields a realization $\Theta=(\lambda G-A,B,C)$ with the extra property that the pencil $\lambda G-A$ has a zero at $\lambda_0\in\mathbb{C}$ if and only if the function W has a pole at λ_0. To be precise, if the function W has no poles in a certain subset V of $\mathbb{C}$, then one may choose realizations $\Theta=(\lambda G-A,B,C)$ with a V-regular pencil $\lambda G-A$.

In the sequel of this manuscript we shall need the following result for rational matrix functions of the form (5.1). Recall that the *Fourier transformation* $\hat{f}$ of and L_1-function f is defined by

$$\hat{f}(\lambda)=\int_{-\infty}^{\infty}e^{i\lambda t}f(t)dt,\quad \lambda\in\mathbb{R}.$$

LEMMA 5.1. *Let W be a rational $m \times m$ matrix function with a realization $\Theta = (\lambda G - A, B, C)$. Assume that the pencil $\lambda G - A$ is $\mathbb{R}$-regular, and let $(E, \Omega, P; P_-, P_+, P_0, P_\infty)$ denote the data for the decomposition of $\lambda G - A$. Then $P_0 = 0$, and W has a unique decomposition $W(\lambda) = q(\lambda) + \hat{k}(\lambda)$, where q is an $m \times m$ matrix polynomial and $\hat{k}$ is the Fourier transform of an L_1-function k. The functions q and k are are given by:*

$$q(\lambda) = I - \sum_{j=0}^{N} \lambda^j CE\Omega^j P_\infty B, \tag{5.2}$$

$$k(t) = \begin{cases} iCEe^{-it\Omega} P_+ B, & t < 0, \\ -iCEe^{-it\Omega} P_- B, & t > 0. \end{cases} \tag{5.3}$$

Here N is the order of the pole at infinity of $(\lambda G - A)^{-1}$.

PROOF. Since W is a rational matrix function, we can write W in a unique way as $W = q + V$, with q a matrix polynomial and V a rational matrix function which is analytic and zero at infinity. To find the above decomposition of W in terms of the chosen realization Θ, write

$$W(\lambda) = I + C(\lambda G - A)^{-1} B = I + CE((\lambda G - A)E)^{-1} B =$$

$$= I + CE((\lambda G - A)E)^{-1} PB + CE((\lambda G - A)E)^{-1} (I - P)B.$$

Using the identities in (2.1) we have

$$W(\lambda) = I + CE(\lambda\Omega(I - P) - I)^{-1} (I - P)B + CE(\lambda - \Omega)^{-1} PB. \tag{5.4}$$

By Lemma 2.1, the matrix $\Omega(I - P)$ is nilpotent of order $N+1$. Hence, the polynomial part of W is given by the first two terms in the right hand side of (5.4), whereas the third term represents the part of W which is analytic and zero at infinity. Since $I - P = P_\infty$, we conclude that q is given by formula (5.2). Furthermore, Lemma 2.2 and the assumption on the pencil $\lambda G - A$ imply that the function k defined by (5.3) is an L_1-function, and that $P_+ + P_- = P$. Applying Fourier transformation, it follows that $V = \hat{k}$, which proves the lemma. □

Henceforth, we shall sometimes refer to the functions q and $\hat{k}$ in (5.2) and (5.3) as the *polynomial part* and the *strictly proper part* of W, respectively. Also, we shall say that a rational matrix function V is *(strictly) proper* if it is analytic at infinity (and $V(\infty) = 0$).

Another important feature of the representation (5.1) is the connection between (5.1) and the invertibility of the function W. Let $W(\lambda)$ be given as in (5.1), and put $A^\times = A - BC$. Then, by Theorem 4.2 in [GK1], the following identity holds:

$$\det W(\lambda) = \frac{\det(\lambda G - A^\times)}{\det(\lambda G - A)}. \tag{5.5}$$

In particular, the function det $W(\lambda)$ is not identically equal to zero if and only if the pencil $\lambda G - A^{\times}$ is regular. Furthermore, if this condition is satisfied, then the inverse function $W(\lambda)^{-1}$ admits a realization $\Theta^{\times} = (\lambda G - A^{\times}, B, -C)$:

$$W(\lambda)^{-1} = I - C(\lambda G - A^{\times})^{-1} B. \tag{5.6}$$

From the above remarks it follows that the conditions (i) and (ii) in Theorem 1.3 are equivalent to the conditions (i') and (ii'), respectively, stated just below Theorem 1.3. Indeed, the equivalence of (i) and (i') is an immediate consequence of identity (5.5). To see that the condition (ii) implies (ii') and vice versa, apply Lemma 5.1 to the realization (5.6).

In the sequel we shall often need information about the interplay between the spectral data of the pencils $\lambda G - A$ and $\lambda G - A^{\times}$ appearing in the realizations (5.1) and (5.6). The next three lemmas present a number of auxiliary results in this direction. In the statements and in the proofs of these lemmas,

$$(E, \Omega, P; P_-, P_+, P_0, P_\infty), \quad (E^{\times}, \Omega^{\times}, P^{\times}; P_-^{\times}, P_+^{\times}, P_0^{\times}, P_\infty^{\times}),$$

denote the data for the decomposition of $\lambda G - A$ and $\lambda G - A^{\times}$, respectively.

LEMMA 5.2. *The space* $\operatorname{Ker} P_- \cap \operatorname{Im} P_-^{\times} \cap \bigcap_{j \geq 0} \operatorname{Ker} CE^{\times}(\Omega^{\times})^j = (0)$.

PROOF. Take a vector $x \in \operatorname{Ker} P_- \cap \operatorname{Im} P_-^{\times} \cap \bigcap_{j \geq 0} \operatorname{Ker} CE^{\times}(\Omega^{\times})^j$. Then $x \in \operatorname{Im} P_-^{\times}$, and hence the identities in (2.1)-(2.3) imply that

$$C(\lambda G - A^{\times})^{-1} x = CE^{\times}((\lambda G - A^{\times})E^{\times})^{-1} P_-^{\times} x = CE^{\times}(\lambda - \Omega^{\times})^{-1} x.$$

Using the Neumann series expansion for the right hand side in the above equality, it follows that

$$C(\lambda G - A^{\times})^{-1} x = \sum_{\nu=0}^{\infty} CE^{\times}(\Omega^{\times})^{\nu} x \lambda^{-\nu-1} \equiv 0. \tag{5.7}$$

Now, use the identity $BC = A - A^{\times} = (\lambda G - A^{\times}) - (\lambda G - A)$ to verify that

$$(\lambda G - A^{\times})^{-1} = (\lambda G - A)^{-1} - (\lambda G - A)^{-1} BC(\lambda G - A^{\times})^{-1}. \tag{5.8}$$

By equation (5.7), the above formula applied to the vector x yields

$$(\lambda G - A^{\times})^{-1} x = (\lambda G - A)^{-1} x.$$

Premultiply this identity by G, and integrate the result over a contour Γ_- which encloses all zeros of the pencils $\lambda G - A$ and $\lambda G - A^{\times}$ in the open lower half plane but excludes all other zeros. This yields $x = P_-^{\times} x = P_- x = 0$. □

The above lemma may be viewed as a generalization of [BGK3], Lemma III.1.1.

Indeed, take $G=I$ and assume that both A and $A^{\times}$ have no spectrum on the real line. Then $P_0=P_{\infty}=0$, and hence Ker $P_-=\text{Im } P_+$. Thus, Lemma 5.2 reduces to Lemma III.1.1 in [BGK3].

LEMMA 5.3. *For $j\geq 1$, the space*

$$\text{Im } (I-\Omega^j(\Omega^{\times})^j)P_{\infty}^{\times}\subset \text{Im } P_{\infty}+\text{Im } (B\ \Omega B\ ..\ \Omega^{j-1}B).$$

PROOF. The proof relies on the formulas

$$P_{\infty}^{\times}=A^{\times}E^{\times}P_{\infty}^{\times}=AE^{\times}P_{\infty}^{\times}-BCE^{\times}P_{\infty}^{\times},$$

and

$$PAE^{\times}P_{\infty}^{\times}=\Omega PE^{-1}E^{\times}P_{\infty}^{\times}=\Omega PGE^{\times}P_{\infty}^{\times}=P\Omega\Omega^{\times}P_{\infty}^{\times}.$$

The above identities are based on the relations (2.1)-(2.3), applied to the pencils $\lambda G-A$ and $\lambda G-A^{\times}$. By combining the above idenities, one obtains

$$P(I-\Omega\Omega^{\times})P_{\infty}^{\times}=PP_{\infty}^{\times}-P\Omega\Omega^{\times}P_{\infty}^{\times}=-PBCE^{\times}P_{\infty}^{\times},$$

which implies the desired inclusion in the lemma for $j=1$. To pass to the case $j>1$, we apply induction to j. Assume that the claim holds for $j=l$. Then, for $j=l+1$ we have

$$P_{\infty}^{\times}-\Omega^{l+1}(\Omega^{\times})^{l+1}P_{\infty}^{\times}=P_{\infty}^{\times}-\Omega^l(\Omega^{\times})^lP_{\infty}^{\times}+\Omega^l(I-\Omega\Omega^{\times})(\Omega^{\times})^lP_{\infty}^{\times}.$$

Thus, using the induction hypotheses

$$\text{Im } (I-\Omega^{l+1}(\Omega^{\times})^{l+1})P_{\infty}^{\times}\subset \text{Im } P_{\infty}+\text{Im } (B\ \Omega B\ ..\ \Omega^{l-1}B)+\text{Im } \Omega^lB,$$

which proves the inclusion for $j=l+1$. □

LEMMA 5.4. *Let r be a nonnegative integer, and assume that the function $W(\lambda)^{-1}\lambda^{-r}$ is proper. Then*

$$PP_{\infty}^{\times}(\Omega^{\times})^{r+j}B=0,\quad j\geq 1. \tag{5.9}$$

PROOF. The assumption on $W(\lambda)^{-1}$ implies that $W(\lambda)^{-1}$ is of the form

$$W(\lambda)^{-1}=I-C(\lambda G-A^{\times})^{-1}P_{\infty}^{\times}B-C(\lambda G-A^{\times})^{-1}P^{\times}B=$$

$$=\lambda^rW_0+\lambda^{r-1}W_1+\lambda^{r-2}W_2+...\qquad(\lambda\to\infty),$$

where W_j, $j=1,2,....$, are certain $m\times m$ matrices. In particular, the polynomial part of $W(\lambda)^{-1}$ has degree at most r. Now, recall formula (5.8) and consider the identity

$$PG(\lambda G-A^{\times})^{-1}P_{\infty}^{\times}B= \tag{5.10}$$

$$=PG(\lambda G-A)^{-1}P_{\infty}^{\times}B-PG(\lambda G-A)^{-1}BC(\lambda G-A^{\times})^{-1}P_{\infty}^{\times}B. \tag{5.10}$$

As $PG(\lambda G-A)^{-1}=P(\lambda-\Omega)^{-1}$ by the identities in Section 2, the right hand side in (5.10) has a pole at infinity of order at most r. Hence, the same has to hold for the left hand side of (5.10). So, the function

$$\lambda^{-r}PGE^{\times}(\lambda GE^{\times}-A^{\times}E^{\times})^{-1}P_{\infty}^{\times}B=-\lambda^{-r}P\Omega^{\times}(\lambda\Omega^{\times}P_{\infty}^{\times}-I)^{-1}P_{\infty}^{\times}B$$

is analytic at infinity. Using the nilpotency of the matrix $\Omega^{\times}P_{\infty}^{\times}$ (cf. Lemma 2.1), this proves the lemma. □

The converse of Lemma 5.4 is not true. To see this, let W be the function

$$W(\lambda)=\begin{pmatrix}1&\lambda\\0&1\end{pmatrix}=\begin{pmatrix}1&0\\0&1\end{pmatrix}+\begin{pmatrix}-1&0\\0&0\end{pmatrix}(\lambda\begin{pmatrix}0&1\\0&0\end{pmatrix}-\begin{pmatrix}1&0\\0&1\end{pmatrix})^{-1}\begin{pmatrix}0&0\\0&1\end{pmatrix}.$$

The corresponding left separating projection $P=0$. So, the equation (5.9) is certainly satisfied for any r. However, the function $W(\lambda)^{-1}\lambda^{j}$ is not analytic at infinity for $j=0,1$, thereby contradicting the assumptions in Lemma 5.4.

6. PROOF OF THEOREM 1.1

In this section we prove Theorem 1.1. For all notations we refer to Section 1.

PROOF OF THEOREM 1.1. Take an element $\phi\in H_s^m(\overline{\mathbb{R}^+})$. Since the pencil $\lambda G-A$ is $\mathbb{R}$-regular, Corollary 3.2 states that the first equation in $(\Sigma]$ with $u=\phi$ has a unique solution in $(\mathcal{S}^n)'$ given by

$$\rho=p(i\frac{d}{dt})B\phi+l*B\phi, \tag{6.1}$$

where the functions p and l are given by

$$p(\lambda)=-\sum_{j=0}^{N}E\Omega^{j}P_{\infty}\lambda^{j}, \tag{6.2}$$

$$l(t)=\begin{cases} iEe^{-it\Omega}P_{+}, & t<0,\\ -iEe^{-it\Omega}P_{-}, & t>0.\end{cases} \tag{6.3}$$

Here N is the order of the pole at infinity of $(\lambda G-A)^{-1}$. A comparison of the formulas (6.2) and (6.3) with those for the functions $q(\lambda)$ and $k(t)$ in Lemma 5.1 yields that

$$Cl(t)B=k(t),\quad I+Cp(\lambda)B=q(\lambda).$$

So, we have, $\phi\in D(T_W)$ and $T_W(\phi)=[f]$ if and only if $C([\rho])+[\phi]=[f]$, which proves the theorem. □

As a preparation to Section 7, we remark that the unique solution $\rho \in (\mathcal{S}^n)'$ to the first equation in $(\Sigma]$ is a distribution in the Sobolev space $H_\mu^n(\mathbb{R})$ for μ sufficiently small. Indeed, this fact is an immediate consequence of the formula in (6.1).

The restriction to the numbers s and r in this section is not essential. With the appropriate modification of the definition of T_W, Theorem 1.1 also holds when s and r are arbitrary real numbers.

7. PROOFS OF THEOREMS 1.2 AND 1.3

In this section we prove Theorems 1.2 and 1.3. We shall freely use the notations introduced in Section 1. In particular, throughout this section $-s$ and $r+s$ are nonnegative integers, the rational function W is given by the realization (1.9), with associated decomposition data (1.12)-(1.13), and $T_W\,(H_s^m(\overline{\mathbb{R}^+}) \to H_{s+r}^m[\mathbb{R}^+])$ is the maximal half-line integro-differential operator with symbol W. We also assume that the pencil $\lambda G - A$ is $\mathbb{R}$-regular, and $A^\times = A - BC$. Furthermore, we shall freely use the equivalence of conditions (i) and (ii) with conditions (i') and (ii') in Section 1 (see Section 5). We start with three lemmas and a proposition.

LEMMA 7.1. *Assume that the conditions (i) and (ii) in Theorem 1.3 are fulfilled. Then the kernel of T_W is given by*

$$\begin{aligned}\operatorname{Ker} T_W = \ &\{(CE^\times e^{-it\Omega^\times}x)_+ - \sum_{j=0}^{-s-1} v_j (i\frac{d}{dt})^j \delta \ \Big| \\ &x \in \operatorname{Im} P_-^\times,\ x + i\sum_{j=0}^{-s-1} \Omega^j B v_j \in \operatorname{Ker} P_-\},\end{aligned} \tag{7.1}$$

$$\dim \operatorname{Ker} T_W = \dim(\operatorname{Im} P_-^\times \cap (\operatorname{Ker} P_- + \operatorname{Im}\mathcal{B})) + \dim(\operatorname{Ker} P_- \cap \operatorname{Im}\mathcal{B}) + \dim \operatorname{Ker}\mathcal{B}. \tag{7.2}$$

PROOF. Let ϕ be an element of $\operatorname{Ker} T_W$. Then, by Theorem 6.1, there exists a distribution ρ in some Sobolev space $H_\mu^n(\mathbb{R})$ such that

$$i\frac{d}{dt}G\rho = A\rho + B\phi, \quad [0] = C[\rho] + [\phi]. \tag{7.3}$$

By applying the mapping q_+ to the first equation in (7.3) we have

$$i\frac{d}{dt}G[\rho] - A[\rho] = B[\phi], \quad [0] = C[\rho] + [\phi].$$

Next, subtract the second equality, premultiplied by B, from the first of the above identities. Thus, one obtains that $[\rho] \in H_\mu^n[\mathbb{R}^+]$ satisfies the equation

$$i\frac{d}{dt}G[\rho] - A^\times[\rho] = [0],$$

and hence, by Corollary 4.2 and assumption (i) of Theorem 1.3, we have

$$[\rho]=[E^{\times}e^{-it\Omega^{\times}}x]$$

for some vector $x\in \operatorname{Im} P_-^{\times}$. As for the distribution ρ itself, this implies the existence of a tempered distribution T_- with support in $(-\infty,0]$ such that

$$\rho=(E^{\times}e^{-it\Omega^{\times}}x)_+ + T_-. \tag{7.4}$$

Furthermore, we have $[\phi]=-C[\rho]=[-CE^{\times}e^{-it\Omega^{\times}}x]$, and hence

$$\phi=-(CE^{\times}e^{-it\Omega^{\times}}x)_+ + T_0 \tag{7.5}$$

for some tempered distribution T_0 with support in $(-\infty,0]$. In fact, as ϕ and the first term in the right hand side of (7.5) are elements of $H_s^m(\overline{\mathbb{R}^+})$, the distribution T_0 has support at the origin and is an element of $H_s^m(\overline{\mathbb{R}^-})$. In other words, using a well-known result from the theory of distributions (cf., *e.g.*, [Tr], Theorem 24.6), T_0 is of the form

$$T_0=v_0\delta+v_1(i\frac{d}{dt})\delta+\ldots+v_{-s-1}(i\frac{d}{dt})^{-s-1}\delta, \tag{7.6}$$

where v_j, $j=0,1,\ldots,-s-1$, are certain vectors in $\mathbb{C}^m$. Now, insert the formulas (7.4)-(7.6) into the first equation in (7.3), and use that $GE^{\times}P^{\times}=P^{\times}$ and $A^{\times}E^{\times}P^{\times}=\Omega^{\times}P^{\times}$ (cf. Section 2). This yields

$$(\Omega^{\times}e^{-it\Omega^{\times}}x)_+ + ix\delta-(AE^{\times}e^{-it\Omega^{\times}}x)_+ + i\frac{d}{dt}GT_- - AT_- =$$

$$=B\phi=((A^{\times}-A)E^{\times}e^{-it\Omega^{\times}}x)_+ + B\sum_{j=0}^{-s-1}v_j(i\frac{d}{dt})^j\delta.$$

The above equality is equivalent to

$$i\frac{d}{dt}GT_- - AT_- = -ix\delta+B\sum_{j=0}^{-s-1}v_j(i\frac{d}{dt})^j\delta. \tag{7.7}$$

As the pencil $\lambda G-A$ is $\mathbb{R}$-regular, Corollary 3.2 states that the distribution T_- is given by

$$T_-=p(i\frac{d}{dt})(-ix\delta+B\sum_{j=0}^{-s-1}v_j(i\frac{d}{dt})^j\delta)+l*(-ix\delta+B\sum_{j=0}^{-s-1}v_j(i\frac{d}{dt})^j\delta), \tag{7.8}$$

where the polynomial p and the integrable function l are as in (6.2) and (6.3). But T_- has its support in $(-\infty,0]$. The same holds for the first term in the right hand side in (7.8). As for the last term in (7.8), observe that

$$l*(\frac{d}{dt})^j\delta = l^{(j)}(t) + \sum_{\nu=0}^{j-1} l^{(j-1-\nu)}(0)(\frac{d}{dt})^{\nu}\delta.$$

Hence, it follows that the third term in (7.8) has its support on the negative halfline if and only if

$$e^{-it\Omega}P_-(-ix + Bv_0 + \Omega Bv_1 + .. + \Omega^{-s-1}Bv_{-s-1}) \equiv 0,$$

or, equivalently, if

$$x + i\sum_{j=0}^{-s-1}\Omega^j Bv_j \in \text{Ker}\, P_-. \tag{7.9}$$

Combining this observation with (7.5) and (7.6), the distribution ϕ is an element of the right hand side in (7.1). Conversely, assume that ϕ is of the form (7.5) with x in the image of $P_-^{\times}$ and T_0 given by (7.6) and (7.9). Define a distribution T_- by (7.8), and let ρ be as in (7.4). Then T_- has its support in $(-\infty,0]$ and satisfies (7.7), the equality $C[\rho]+[\phi]=0$ holds and a computation as above shows that the first equation in (7.3) is satisfied. But then, Theorem 6.1 shows that $\phi \in \text{Ker}\, T_W$.

Next, let us prove (7.2). Introduce an $2n \times (n-ms)$ matrix Γ by

$$\Gamma = \begin{bmatrix} I_n & 0 \\ I_n & i\mathcal{B} \end{bmatrix} : \mathbb{C}^n \times \mathbb{C}^{-ms} \to \mathbb{C}^n \times \mathbb{C}^n,$$

and denote by $\mathbf{v}$ the column vector containing as elements the vectors v_0 up to v_{-s-1}. More precisely, $\mathbf{v} = (v_0^T \ v_1^T \ \cdots \ v_{-s-1}^T)^T$. In terms of the matrix Γ the kernel of T_W consists of those $\phi \in H_s^m(\overline{\mathbb{R}^+})$ of the form (7.5) for which the vectors x and $\mathbf{v}$ satisfy the condition

$$\Gamma\begin{bmatrix} x \\ \mathbf{v} \end{bmatrix} \in \text{Im}\, P_-^{\times} \times \text{Ker}\, P_-. \tag{7.10}$$

We claim that the correspondence between the vectors x and $\mathbf{v}$ satisfying (7.10) and elements $\phi \in \text{Ker}\, T_W$ is one-one. Indeed, take vectors $x \in \text{Im}\, P_-^{\times}$ and v_j, $j=0,1,..,-s-1$, in $\mathbb{C}^m$ satisfying (7.10), and assume that the distribution ϕ in (7.5)-(7.6) is the zero distribution. Then it is immediately clear that $v_j = 0$ for $j=0,1,..,-s-1$, and that $CE^{\times}e^{-it\Omega^{\times}}x \equiv 0$ for $t>0$. Hence $\mathbf{v}=0$, and thus x is contained in $\text{Im}\, P_-^{\times} \cap \text{Ker}\, P_-$ because of (7.10). Now apply Lemma 5.2 to conclude that $x=0$. This proves the claim.

By the results of the previous paragraph we have

$$\dim \text{Ker}\, T_W = \dim \Gamma^{-1}[\text{Im}\, P_-^{\times} \times \text{Ker}\, P_-] = \dim \Gamma^{-1}[(\text{Im}\, P_-^{\times} \times \text{Ker}\, P_-) \cap \text{Im}\, \Gamma] =$$

$$= \dim \text{Ker}\, \Gamma + \dim\big((\text{Im}\, P_-^{\times} \times \text{Ker}\, P_-) \cap \text{Im}\, \Gamma\big).$$

Consider the second term in the above expression. If the vector $\Gamma\begin{bmatrix} x \\ \mathbf{v} \end{bmatrix}$ is contained in the space

$\operatorname{Im} P_-^\times \times \operatorname{Ker} P_-$, we have

$$x \in \operatorname{Im} P_-^\times \cap (\operatorname{Ker} P_- + \operatorname{Im} \mathcal{B}) =: \mathcal{V}.$$

Conversely, if $x \in \mathcal{V}$, then there exist vectors v_0 up to v_{-s-1} such that $\Gamma \begin{pmatrix} x \\ \mathbf{v} \end{pmatrix}$ is an element of $\operatorname{Im} P_-^\times \times \operatorname{Ker} P_-$, and the number of possible linearly independent vectors $\Gamma \begin{pmatrix} x \\ \mathbf{v} \end{pmatrix}$ equals the dimension of $\operatorname{Im} \mathcal{B} \cap \operatorname{Ker} P_-$. Thus, we derived that

$$\dim (\operatorname{Im} P_-^\times \times \operatorname{Ker} P_-) \cap \operatorname{Im} \Gamma = \dim \operatorname{Im} P_-^\times \cap (\operatorname{Ker} P_- + \operatorname{Im} \mathcal{B}) + \dim (\operatorname{Ker} P_- \cap \operatorname{Im} \mathcal{B}).$$

As $\operatorname{Ker} \Gamma = (0) \times \operatorname{Ker} \mathcal{B}$, this establishes formula (7.2). □

LEMMA 7.2. *Assume that the conditions (i) and (ii) in Theorem 1.3 are fulfilled. Then the image of T_W is given by*

$$\operatorname{Im} T_W = \{q_+(f) \mid f \in H_{s+r}^m(\mathbb{R}),\ R_{r,s}(Bf) \in \operatorname{Im} P_-^\times + \operatorname{Ker} P_- + \operatorname{Im} \mathcal{B}\}, \tag{7.11}$$

$$\operatorname{codim} \operatorname{Im} T_W = \operatorname{codim} (\operatorname{Im} P_-^\times + \operatorname{Ker} P_- + \operatorname{Im} \mathcal{B}). \tag{7.12}$$

Furthermore, for a function $[f] \in \operatorname{Im} T_W$ the general solution to the equation $T_W(\phi) = [f]$ in the space $H_s^m(\overline{\mathbb{R}^+})$ is given by

$$\begin{aligned}
\phi &= f_+ + \sum_{j=0}^{r+s} CE^\times (i\Omega^\times)^j P_\infty^\times B(f^{(j)})_+ + \\
&+ \sum_{j=r+s+1}^{r} CE^\times (i\Omega^\times)^j P_\infty^\times B(\frac{d}{dt})^{j-r-s}((f^{(r+s)})_+) + \\
&+ (\int_0^\infty k^\times(t-s) f_+(s) ds - CE^\times e^{-it\Omega^\times} x)_+ + \sum_{j=0}^{-s-1} v_j (i\frac{d}{dt})^j \delta + \\
&+ \sum_{j=0}^{-s-1} \sum_{\nu=j}^{r+s+j-1} CE^\times (i\Omega^\times)^{\nu+1} P_\infty^\times B f^{(\nu-j)}(0) (\frac{d}{dt})^j \delta,
\end{aligned} \tag{7.13}$$

where $f \in H_{s+r}^m(\mathbb{R})$ is any representative of $[f]$, and $x \in \operatorname{Im} P_-^\times$ and $v_j \in \mathbb{C}^m$, $j = 0, 1, \ldots, -s-1$, are vectors such that

$$R_{r,s}(Bf) + x + i(Bv_0 + \Omega B v_1 + \cdots + \Omega^{-s-1} B v_{-s-1}) \in \operatorname{Ker} P_-. \tag{7.14}$$

Before proving Lemma 7.2 we need the following result. Recall that $\mathcal{S}^m$ is the space of rapidly decreasing functions with values in $\mathbb{C}^m$.

LEMMA 7.3. *Define the mapping $R_{\infty,s} : \mathcal{S}^n \to \mathbb{C}^n$ by*

$$R_{\infty,s}(f) = i\Omega^{-s} (\Omega^\times)^{-s} \sum_{j=0}^{\infty} (i\Omega^\times)^{j+1} P_\infty^\times f^{(j)}(0) + i\int_0^\infty e^{is\Omega^\times} P_+^\times f(s) ds.$$

Then Ker $P_- + \operatorname{Im} P_-^\times + \operatorname{Im} P_0^\times + \operatorname{Im} R_{\infty,s} \; B + \operatorname{Im} \mathcal{B} = \mathbb{C}^n$.

PROOF. Recall from Lemma 2.1 that the matrix $\Omega^\times P_\infty^\times$ is nilpotent. Hence, the mapping $R_{\infty,s}$ is well-defined. Let v be a vector in the image of P_-, and choose a function $f \in \mathcal{S}^m$ such that $f(t) = CEe^{-it\Omega} v$ for $t > 0$. First, we prove the following two identities:

$$i\int_0^\infty e^{is\Omega^\times} P_+^\times Bf(s)ds = P_+^\times v, \tag{7.15}$$

$$i\sum_{j=0}^\infty (i\Omega^\times)^{j+1} P_\infty^\times Bf^{(j)}(0) = P_\infty^\times v. \tag{7.16}$$

To verify (7.15), use the identities $AEP = \Omega P$ and $P^\times A^\times E^\times = \Omega^\times P^\times$ to check that

$$i\int_0^\infty e^{is\Omega^\times} P_+^\times Bf(s)ds = i\int_0^\infty e^{is\Omega^\times} P_+^\times (A - A^\times)Ee^{-is\Omega} vds =$$

$$= i\int_0^\infty e^{is\Omega^\times} P_+^\times \Omega e^{-is\Omega} vds - i\int_0^\infty e^{is\Omega^\times} P_+^\times \Omega^\times (E^\times)^{-1} Ee^{-is\Omega} vds.$$

Rewrite the first integral in the previous formula as

$$i\int_0^\infty e^{is\Omega^\times} P_+^\times \Omega e^{-is\Omega} vds = -e^{is\Omega^\times} P_+^\times e^{-is\Omega} v \Big|_0^\infty + i\Omega^\times \int_0^\infty e^{is\Omega^\times} P_+^\times e^{-is\Omega} vds =$$

$$= P_+^\times v + i\int_0^\infty e^{is\Omega^\times} P_+^\times \Omega^\times e^{-is\Omega} vds.$$

Since $P^\times GE^\times = P^\times$ and $GEP = P$ (cf. the formulas (2.1)-(2.3)), we have $P^\times (E^\times)^{-1} EP = P^\times P$, which proves (7.15). Next, to verify (7.16), recall the formulas $AEP = \Omega P$ and $P_\infty^\times A^\times E^\times = P_\infty^\times$, and check that

$$i\sum_{j=0}^\infty (i\Omega^\times)^{j+1} P_\infty^\times Bf^{(j)}(0) = i\sum_{j=0}^\infty (i\Omega^\times)^{j+1} P_\infty^\times (A - A^\times)E(-i\Omega)^j v =$$

$$= -\sum_{j=0}^\infty (i\Omega^\times)^{j+1} P_\infty^\times (-i\Omega)^{j+1} v - i\sum_{j=0}^\infty (i\Omega^\times)^{j+1} P_\infty^\times (E^\times)^{-1} E(-i\Omega)^j v.$$

To complete the proof of (7.16), it now suffices to compute the right hand side in the previous expression by employing the identities

$$\Omega^\times P_\infty^\times (E^\times)^{-1} EP = P_\infty^\times GE^\times (E^\times)^{-1} EP = P_\infty^\times P$$

(cf. Section 2), which yields the desired identity. Now, to conclude the proof, combine the identities in (7.15) and (7.16) and verify that $R_{\infty,s}(Bf) = P_+^\times v + \Omega^{-s}(\Omega^\times)^{-s} P_\infty^\times v$. Hence, we may write the vector v as

$$v=P_-^\times v+P_0^\times v+R_{\infty,s}(Bf)+(P_\infty^\times v-\Omega^{-s}(\Omega^\times)^{-s}P_\infty^\times v).$$

By Lemma 5.3, the last term in the previous expression is an element of Ker P_- + Im $\mathcal{B}$. As the same holds for vectors v in Ker P_-, this finishes the proof. □

The above lemma may be viewed as a generalization of Lemma II.1.1 in [BGK3]. Indeed, take $G=I$, and assume that both A and $A^\times$ have no spectrum on the real line. Then $P_0=P_\infty=0$, and hence Ker P_- = Im P_+. Furthermore, we have $P_\infty^\times=0=P_0^\times$ and $P_-^\times+P_+^\times=I$, and so

$$\text{Im}\,R_{\infty,s}\ B=\text{Im}\,[P_+^\times B\quad P_+^\times\Omega^\times B\ \dots\ P_+^\times(\Omega^\times)^nB]\subset\text{Im}\,P_-^\times+\text{Im}\,[B\ \ \Omega^\times B\ \dots\ (\Omega^\times)^nB].$$

It follows that Im $P_-^\times$ + Im P_+ + Im $[B\ \ \Omega^\times B\ \dots\ (\Omega^\times)^nB]=\mathbb{C}^n$, which is [BGK3], Lemma II.1.1.

PROOF OF LEMMA 7.2. The proof is divided into three parts.

Part (A). Let $[f]\in$ Im T_W. In this part we prove that $[f]$ is contained in the right hand side of (7.11), and that any solution $\phi\in H_s^m(\overline{\mathbb{R}^+})$ of the equation $T_W(\phi)=[f]$ is of the form (7.13)-(7.14). Throughout Part (A), the expression f denotes a representative of the class $[f]\in H_{s+r}^m[\mathbb{R}^+]$ in the space $H_{s+r}^m(\mathbb{R})$.

To start the proof, let $[f]$, f and ϕ be as above. Then, by Theorem 6.1, there exists a distribution ρ satisfying the equalities

$$i\frac{d}{dt}G\rho-A\rho=B\phi,\quad [f]=C[\rho]+[\phi]. \tag{7.17}$$

By passing to the inverse system, one obtains

$$i\frac{d}{dt}G[\rho]=A^\times[\rho]+B[f],\quad [\phi]=-C[\rho]+[f].$$

Now, apply Theorem 4.1 to the first equation above. Using assumption (i) of Theorem 1.3, it follows that there exists a vector $x\in$ Im $P_-^\times$ such that

$$[\rho]=\Big[p^\times(i\frac{d}{dt})Bf_++\int_0^\infty l^\times(t-s)Bf_+(s)ds\Big]+\Big[E^\times e^{-it\Omega^\times}x\Big], \tag{7.18}$$

where

$$p^\times(\lambda)=-\sum_{j=0}^{N^\times}E^\times(\Omega^\times)^jP_\infty^\times\lambda^j,$$

$$l^{\times}(t)=\begin{cases} iE^{\times}e^{-it\Omega^{\times}}P_{+}^{\times}, & t<0,\\ -iE^{\times}e^{-it\Omega^{\times}}P_{-}^{\times}, & t>0,\end{cases}$$

and $N^{\times}$ is the order of the pole at infinity of $(\lambda G-A^{\times})^{-1}$. Hence, the second equality in (7.17) implies that the distribution ϕ belongs to the coset

$$[\phi]=\left[q^{\times}(i\frac{d}{dt})f_{+}+\int_{0}^{\infty}k^{\times}(t-s)f_{+}(s)ds\right]-\left[CE^{\times}e^{-it\Omega^{\times}}x\right], \tag{7.19}$$

with $k^{\times}$ the function given in Theorem 1.2 and $q^{\times}$ the polynomial

$$q^{\times}(\lambda)=I+\sum_{j=0}^{r}CE^{\times}(\Omega^{\times})^{j}P_{\infty}^{\times}B\lambda^{j}.$$

Here the summation stops at the level $j=r$ because of assumption (ii) of Theorem 1.3.

Next, we establish the formulas (7.13) and (7.14). This part of the proof is divided into three steps.

Step (i). As a preparation we first prove that $q^{\times}(i\frac{d}{dt})f_{+}=H+R$ for distributions $H\in H_{s}^{m}(\overline{\mathbb{R}^{+}})$ and $R\in(\mathcal{S}^{m})'(\overline{\mathbb{R}^{-}})$ given by

$$\begin{aligned} H= {} & f_{+}+\sum_{j=0}^{r+s}CE^{\times}(i\Omega^{\times})^{j}P_{\infty}^{\times}B(f^{(j)})_{+}+\\ & +\sum_{j=r+s+1}^{r}CE^{\times}(i\Omega^{\times})^{j}P_{\infty}^{\times}B(\frac{d}{dt})^{j-r-s}((f^{(r+s)})_{+})+\\ & +\sum_{\mu=0}^{-s-1}\sum_{j=\mu+1}^{r+s+\mu}CE^{\times}(i\Omega^{\times})^{j}P_{\infty}^{\times}Bf^{(j-1-\mu)}(0)(\frac{d}{dt})^{\mu}\delta, \end{aligned} \tag{7.20}$$

$$R=\sum_{\mu=-s}^{r-1}\sum_{j=\mu+1}^{r}CE^{\times}(i\Omega^{\times})^{j}P_{\infty}^{\times}Bf^{(j-1-\mu)}(0)(\frac{d}{dt})^{\mu}\delta. \tag{7.21}$$

To establish (7.20) and (7.21), first observe that

$$(\frac{d}{dt})^{j}f_{+}=(f^{(j)})_{+}+\sum_{\nu=0}^{j-1}f^{(\nu)}(0)(\frac{d}{dt})^{j-1-\nu}\delta,\quad j=0,1,\ldots,r+s.$$

For integers $j\geq r+s$ we have

$$(\frac{d}{dt})^{j}f_{+}=(\frac{d}{dt})^{j-(r+s)}(f^{(r+s)})_{+}+\sum_{\nu=0}^{r+s-1}f^{(\nu)}(0)(\frac{d}{dt})^{j-1-\nu}\delta.$$

Combining these formulas, a straightforward computation shows that

$$q^{\times}(i\frac{d}{dt})f_{+}=f_{+}+\sum_{j=0}^{r}CE^{\times}(i\Omega^{\times})^{j}P_{\infty}^{\times}B(\frac{d}{dt})^{j}f_{+}=$$

$$=f_+ + \sum_{j=0}^{r+s} CE^\times (i\Omega^\times)^j P_\infty^\times B\Big((f^{(j)})_+ + \sum_{\nu=0}^{j-1} f^{(\nu)}(0)(\frac{d}{dt})^{j-1-\nu}\delta\Big) +$$

$$+ \sum_{j=r+s+1}^{r} CE^\times (i\Omega^\times)^j P_\infty^\times B\Big((\frac{d}{dt})^{j-r-s}((f^{(r+s)})_+) + \sum_{\nu=0}^{r+s-1} f^{(\nu)}(0)(\frac{d}{dt})^{j-1-\nu}\delta\Big) =$$

$$=f_+ + \sum_{j=0}^{r+s} CE^\times (i\Omega^\times)^j P_\infty^\times B(f^{(j)})_+ + \tag{7.22}$$

$$+ \sum_{j=r+s+1}^{r} CE^\times (i\Omega^\times)^j P_\infty^\times B(\frac{d}{dt})^{j-r-s}((f^{(r+s)})_+) +$$

$$+ \sum_{j=0}^{r+s} CE^\times (i\Omega^\times)^j P_\infty^\times B \sum_{\nu=0}^{j-1} f^{(\nu)}(0)(\frac{d}{dt})^{j-1-\nu}\delta + \tag{7.23}$$

$$+ \sum_{j=r+s+1}^{r} CE^\times (i\Omega^\times)^j P_\infty^\times B \sum_{\nu=0}^{r+s-1} f^{(\nu)}(0)(\frac{d}{dt})^{j-1-\nu}\delta. \tag{7.24}$$

The terms in (7.22) belong to $H_s^m(\overline{\mathbb{R}^+})$. To analyse the terms in (7.23) and (7.24), perform the substitution $\mu := j-1-\nu$, recall that $-s \le r$ by the assumption on the integers r and s, and separate the terms containing derivatives of δ of order smaller than $-s$ from those with derivatives of order greater than or equal to $-s$. Thus, the sum of the terms in (7.23) and (7.24) may be rewritten as

$$\sum_{\mu=0}^{-s-1} \sum_{j=\mu+1}^{r+s+\mu} CE^\times (i\Omega^\times)^j P_\infty^\times B f^{(j-1-\mu)}(0)(\frac{d}{dt})^\mu \delta +$$

$$+ \sum_{\mu=-s}^{r-1} \sum_{j=\mu+1}^{r} CE^\times (i\Omega^\times)^j P_\infty^\times B f^{(j-1-\mu)}(0)(\frac{d}{dt})^\mu \delta.$$

The second double sum in the previous formula equals R, while the first double sum and (7.22) together equal H. It follows that $q^\times(i\frac{d}{dt})f_+$ has the desired representation.

Step (ii). Let the function $k^\times$ and the vector x be as in (7.19), and introduce a distribution ϕ_1 by

$$\phi_1 = H + (\int_0^\infty k^\times(t-s) f_+(s)ds)_+ - (CE^\times e^{-it\Omega^\times} x)_+. \tag{7.25}$$

In this step we show that

$$\phi = \phi_1 + v_0\delta + v_1(i\frac{d}{dt})\delta + \cdots + v_{-s-1}(i\frac{d}{dt})^{-s-1}\delta \tag{7.26}$$

for certain vectors v_j, $j=0,1,\dots,-s-1$, in $\mathbb{C}^m$. Indeed, from (7.20) and (7.25) we see that ϕ_1 is an element of $H_s^m(\overline{\mathbb{R}^+})$. Furthermore, because the distribution R in (7.21) has its support in $\overline{\mathbb{R}^-}$, we have

$$[H]=[q^{\times}(i\tfrac{d}{dt})f_{+}],$$

and hence the distribution ϕ_1 satisfies $[\phi_1]=[\phi]$ (cf. (7.25) and (7.19)). But then, the support of $\phi-\phi_1$ is contained in $(-\infty,0]$. On the other hand $\phi-\phi_1$ is contained in $H_s^m(\overline{\mathbb{R}^+})$. Combining these observations, equality (7.26) follows.

Step (iii). In the third step we show that the vectors v_j in (7.26) have to satisfy the condition in (7.14). To prove this, we first define a distribution ρ_1 by

$$\rho_1=p^{\times}(i\frac{d}{dt})Bf_{+}+\Big(\int_0^{\infty}l^{\times}(t-s)Bf_{+}(s)ds\Big)_{+}+\big(E^{\times}e^{-it\Omega^{\times}}x\big)_{+}. \tag{7.27}$$

Here the functions $p^{\times}$ and $l^{\times}$ and the vector $x\in \operatorname{Im} P_{-}^{\times}$ are as in (7.18). Then, as $[\rho]=[\rho_1]$, the distribution ρ is of the form $\rho=\rho_1+T_{-}$ for some tempered distribution T_{-} with support in $(-\infty,0]$. Insert this formula into the first equality in (7.17). This yields

$$B\phi=i\frac{d}{dt}G\rho_1-A^{\times}\rho_1-BC\rho_1+i\frac{d}{dt}GT_{-}-AT_{-}.$$

Next, recall from Corollary 3.2 that

$$p^{\times}(i\frac{d}{dt})Bf_{+}+\int_0^{\infty}l^{\times}(t-s)Bf_{+}(s)ds$$

is the unique solution to the equation $i\frac{d}{dt}G\tilde{\rho}-A^{\times}\tilde{\rho}=Bf_{+}$. Using this fact as well as the formula in (7.27), we obtain the identity

$$B\phi=Bf_{+}+(ix-\int_0^{\infty}e^{is\Omega^{\times}}P_{+}^{\times}Bf(s)ds)\delta-BC\rho_1+i\frac{d}{dt}GT_{-}-AT_{-}.$$

Furthermore, observe that $\phi_1+R=f_{+}-C\rho_1$. Hence, employing (7.26), we arrive at the following equality for the distribution T_{-}:

$$i\frac{d}{dt}GT_{-}-AT_{-}=(-ix+\int_0^{\infty}e^{is\Omega^{\times}}P_{+}^{\times}Bf(s)ds)\delta+\sum_{j=0}^{-s-1}Bv_j(i\frac{d}{dt})^j\delta-BR. \tag{7.28}$$

Now, recall that T_{-} has its support in $(-\infty,0]$, and repeat a reasoning in the proof of Lemma 7.1 (see the formulas in (7.7)-(7.9)). It follows that

$$-ix+\sum_{j=0}^{-s-1}\Omega^jBv_j+\int_0^{\infty}e^{is\Omega^{\times}}P_{+}^{\times}Bf(s)ds-\sum_{\mu=-s}^{r-1}(-i\Omega)^{\mu}Bw_{\mu}\in \operatorname{Ker} P_{-}, \tag{7.29}$$

where

$$w_{\mu} = \sum_{j=\mu+1}^{r} CE^{\times}(i\Omega^{\times})^{j} P_{\infty}^{\times} Bf^{(j-1-\mu)}(0).$$

Premultiply the last term in (7.29) by P_{-}. This yields, using the equality $A^{\times}E^{\times}P_{\infty}^{\times} = P_{\infty}^{\times}$,

$$P_{-}(\sum_{\mu=-s}^{r-1}(-i\Omega)^{\mu}Bw_{\mu}) = \sum_{\mu=-s}^{r-1}\sum_{j=\mu+1}^{r} P_{-}(-i\Omega)^{\mu}BCE^{\times}(i\Omega^{\times})^{j}P_{\infty}^{\times}Bf^{(j-1-\mu)}(0) =$$

$$= \sum_{\mu=-s}^{r-1}\sum_{j=\mu+1}^{r} P_{-}(-i\Omega)^{\mu}AE^{\times}(i\Omega^{\times})^{j}P_{\infty}^{\times}Bf^{(j-1-\mu)}(0) + \tag{7.30}$$

$$-\sum_{\mu=-s}^{r-1}\sum_{j=\mu+1}^{r} P_{-}(-i\Omega)^{\mu}(i\Omega^{\times})^{j}P_{\infty}^{\times}Bf^{(j-1-\mu)}(0). \tag{7.31}$$

As $PA = \Omega PE^{-1} = \Omega PG$ and $GE^{\times}P_{\infty}^{\times} = \Omega^{\times}P_{\infty}^{\times}$ (cf. the identities (2.1)-(2.3)), the expression in (7.30) equals

$$\sum_{\mu=-s}^{r-1}\sum_{j=\mu+1}^{r} iP_{-}(-i\Omega)^{\mu+1}GE^{\times}(i\Omega^{\times})^{j}P_{\infty}^{\times}Bf^{(j-1-\mu)}(0) =$$

$$= \sum_{\mu=-s}^{r-1}\sum_{j=\mu+1}^{r} P_{-}(-i\Omega)^{\mu+1}(i\Omega^{\times})^{j+1}P_{\infty}^{\times}Bf^{(j-1-\mu)}(0).$$

Hence, the expressions in (7.30) and (7.31) add up to

$$\sum_{\mu=-s+1}^{r} P_{-}(-i\Omega)^{\mu}(i\Omega^{\times})^{r+1}P_{\infty}^{\times}Bf^{(r-\mu)}(0) - \sum_{j=-s+1}^{r} P_{-}(-i\Omega)^{-s}(i\Omega^{\times})^{j}P_{\infty}^{\times}Bf^{(j-1+s)}(0).$$

Because of the second assumption of Theorem 1.3, the first sum vanishes. Indeed, by Lemma 5.4, the matrix $(-i\Omega)^{\mu}P_{-}P_{\infty}^{\times}(\Omega^{\times})^{r+1}B = 0$. The second sum equals

$$-\sum_{j=0}^{r+s-1} P_{-}(-i\Omega)^{-s}(i\Omega^{\times})^{-s+j+1}P_{\infty}^{\times}Bf^{(j)}(0).$$

Thus, as a condition on the vectors v_j, we derive from (7.29) that

$$\int_0^{\infty} e^{is\Omega^{\times}}P_{+}^{\times}Bf(s)ds + (-i\Omega)^{-s}(i\Omega^{\times})^{-s}\sum_{j=0}^{r+s-1}(i\Omega^{\times})^{j+1}P_{\infty}^{\times}Bf^{(j)}(0) + \sum_{j=0}^{-s-1}\Omega^{j}Bv_j - ix \in \operatorname{Ker} P_{-},$$

which is condition (7.11).

From the results obtained in Step (i) up to Step (iii), and in particular the identities in (7.20) and (7.25)-(7.27), it follows that the solution ϕ is of the form (7.13)-(7.14). Furthermore, we proved that the representative f of $[f] \in \operatorname{Im} T_W$ (and hence any representative of $[f]$) satisfies condition (7.14), and thus $[f]$ is contained in the right hand side of (7.11). This concludes

Part (A).

Part (B). Assume that the function $f \in H^m_{s+r}(\mathbb{R})$ and the vectors $x \in \operatorname{Im} P^\times_-$ and $v_j \in \mathbb{C}^m$, $j=0,1,...,-s-1$, satisfy the condition in (7.14). In this part of the proof we show that, under this assumption, the distribution ϕ in (7.13) solves the equation $T_W(\phi)=q_+(f)$. In particular, we prove that any distribution $q_+(f)$ contained in the right hand side of (7.11) is an element of $\operatorname{Im} T_W$.

To prove our claim, let f, x, and v_j be as above. Define a distribution ϕ by (7.26) with ϕ_1 as in (7.20) and (7.25). Then $\phi \in H^m_s(\overline{\mathbb{R}^+})$, ϕ is given by (7.13), and $[\phi]=[\phi_1]$. Next, define distributions R and ρ_1 as in (7.21) and (7.27), let T_- be the unique solution to equation (7.28) (which exists since $\lambda G - A$ is $\mathbb{R}$-regular; see Corollary 3.2), and put $\rho=\rho_1+T_-$. Then, since $\operatorname{supp} R \subset (-\infty,0]$, the calculations in Step (i) show that $[\phi]=[\phi_1]$ is given by (7.19). Furthermore, using the expression for the solution of (7.28) given in Corollary 3.2 and condition (7.14), it follows that $T_- \in (\mathcal{S}^n)'$ has its support in $(-\infty,0]$. Hence, $[\rho]=[\rho_1]$ is as in (7.18), and a straightforward computation shows that

$$[f]-C[\rho]=[f_+]-C[\rho]=[\phi].$$

Moreover, employing the calculations in Step (iii),

$$i\frac{d}{dt}G\rho - A\rho = i\frac{d}{dt}G\rho_1 - A\rho_1 + i\frac{d}{dt}GT_- - AT_- =$$

$$= Bf_+ - BC\rho_1 + \sum_{j=0}^{-s-1} Bv_j\left(i\frac{d}{dt}\right)^j\delta - BR = B\left(\phi_1 + \sum_{j=0}^{-s-1} v_j\left(i\frac{d}{dt}\right)^j\delta\right) = B\phi.$$

In other words, the distributions ϕ, ρ and $[f]$ satisfy the equations in (7.17). But then, Theorem 6.1 shows that $T_W(\phi)=[f]$.

Part (C). In this part we verify equality (7.12). Introduce a mapping

$$\frac{H^m_{s+r}[\mathbb{R}^+]}{\operatorname{Im} T_W} \to \frac{\mathbb{C}^n}{\operatorname{Ker} P_- + \operatorname{Im} P^\times_- + \operatorname{Im} \mathcal{B}},$$

by

$$q_+(f) + \operatorname{Im} T_W \;\; R_{r,s}(Bf) + \left(\operatorname{Ker} P_- + \operatorname{Im} P^\times_- + \operatorname{Im} \mathcal{B}\right).$$

By equality (7.11) the above mapping is well-defined and injective. We claim that it is surjective as well, or, equivalently, we claim that

$$\operatorname{Im} R_{r,s} B + \operatorname{Ker} P_- + \operatorname{Im} P^\times_- + \operatorname{Im} \mathcal{B} = \mathbb{C}^n.$$

To establish the above identity, combine assumption (ii) of Theorem 1.3 and Lemma 5.4 to check that

$$P_-[\operatorname{Im} R_{\infty,s}\ B]\subset P_-[\operatorname{Im} R_{r,s}\ B].$$

An application of Lemma 7.3 now completes the proof. □

PROPOSITION 7.4. *Let X be a Banach space, and let M be a closed subspace of X of finite codimension. If D is a dense subspace of X that contains M, then $D = X$.*

PROOF. This lemma immediately follows from (and, in fact, is equivalent to) the well-known fact that any subspace D that contains a closed subspace of finite codimension, is closed itself. □

PROOF OF THEOREM 1.3. The proof is divided in three steps. In Step (i) we show that T_W cannot be Fredholm if $\det W(\lambda)$ is identically equal to zero. In the second step we assume the operator T_W to be Fredholm and derive the properties (i) and (ii) mentioned in the theorem. Step (iii) concerns the reverse implication and the statements in (1.16) up to (1.22).

Step (i). Suppose that $\det W(\lambda)\equiv 0$. Apply the Smith-McMillan form and write the function W as $W(\lambda)=E(\lambda)D(\lambda)F(\lambda)$, where E and F are $m\times m$ matrix polynomials with constant nonzero determinants and D is a rational matrix function of the form

$$D(\lambda)=\begin{bmatrix} D_1(\lambda) & 0\\ 0 & 0\end{bmatrix},$$

with D_1 a diagonal matrix. By the assumption on $\det W$, the size of D_1 is $l\times l$ for some $l<m$. Now, recall that F^{-1} is a matrix polynomial as well, and define a subspace $\mathcal{V}_0$ by

$$\mathcal{V}_0=\{F^{-1}(i\frac{d}{dt})(0,0,\ldots,\phi)^T \mid \phi\in\mathcal{S}(\overline{\mathbb{R}^+})\}.$$

Here $\mathcal{S}(\overline{\mathbb{R}^+})$ stands for the rapidly decreasing functions in $\mathcal{S}$ with support in $\overline{\mathbb{R}^+}$. Then, using the relations between differentiation, convolution and Fourier transformation in $\mathcal{S}'$, a straightforward computation shows that $\mathcal{V}_0\subset\operatorname{Ker} T_W$. Furthermore, the subspace $\mathcal{V}_0$ is infinite dimensional. Indeed, as

$$F(i\frac{d}{dt})F^{-1}(i\frac{d}{dt})\rho=\rho,\quad \rho\in(\mathcal{S}^m)',$$

the mapping $\rho\ F^{-1}(i\frac{d}{dt})\rho$ acting from $(\mathcal{S}^m)'$ to $(\mathcal{S}^m)'$ is injective. But then the space $\operatorname{Ker} T_W$ is infinite dimensional, and thus T_W cannot be a Fredholm operator.

Step (ii). Assume that the operator T_W is Fredholm. Then, by Step (i), the determinant of $W(\lambda)$ does not vanish identically. Therefore, using identity (5.5), the pencil $\lambda G - A^\times$ is regular. In particular, the decomposition data (1.13) for $\lambda G - A^{\times}$ are well-defined.

Next, we reduce the problem to the case $r = s = 0$. Define a function $\tilde{W}$ by

$$\tilde{W}(\lambda) = (\lambda + i)^{r+s} W(\lambda)(\lambda - i)^{-s},$$

and let $T_{\tilde{W}}\ (L_2^m(\overline{\mathbb{R}^+}) \to L_2^m[\mathbb{R}^+])$ be the maximal half-line integro-differential operator with symbol $\tilde{W}$. Here the expressions $L_2^n(\overline{\mathbb{R}^+})$ and $L_2^n[\mathbb{R}^+]$ stand for the spaces $H_0^n(\overline{\mathbb{R}^+})$ and $H_0^n[\mathbb{R}^+]$, respectively. The operators T_W and $T_{\tilde{W}}$ are linked by the following relations. A function $\phi \in L_2^m(\overline{\mathbb{R}^+})$ is an element of $D(T_{\tilde{W}})$ if and only if $(i\frac{d}{dt} - i)^{-s}\phi \in D(T_W)$, and in that case

$$T_{\tilde{W}}(\phi) = (i\frac{d}{dt} + i)^{r+s} T_W((i\frac{d}{dt} - i)^{-s}\phi).$$

Observe that the operators

$$\phi\ (i\frac{d}{dt} - i)^{-s} \in H_s^m(\overline{\mathbb{R}^+}), \quad \phi \in L_2^m(\overline{\mathbb{R}^+}),$$

$$[f]\ (i\frac{d}{dt} + i)^{r+s}[f] \in L_2^m[\mathbb{R}^+], \quad [f] \in H_{s+r}^m[\mathbb{R}^+],$$

are invertible bounded linear mappings. Hence, along with the operator T_W, the operator $T_{\tilde{W}}$ is Fredholm. Once it is proved that this implies that det $\tilde{W}(\lambda) \neq 0$ for $\lambda \in \mathbb{R}$ and that the function $\tilde{W}^{-1}(.)$ is analytic at infinity, the properties (i') and (ii'), and hence (i) and (ii), for $W(\lambda)$ follow immediately.

From now on we assume the integers r and s to be zero. First, we introduce two linear subspaces D_1 and D_2 of $L_2^m[\mathbb{R}^+]$, namely:

$$D_1 = \{q_+(f) \mid f \in L_2^m(\mathbb{R}), \exists x \in \text{Im } P_0^{\times} : \left(e^{-it\Omega^{\times}}(x - i\int_0^t e^{is\Omega^{\times}} P_0^{\times} Bf(s)ds)\right)_+ \in L_2^n(\overline{\mathbb{R}^+})\},$$

$$D_2 = \{[f] \in L_2^m[\mathbb{R}^+] \mid \sum_{j=0}^{N^{\times}} CE^{\times}(i\Omega^{\times})^j P_{\infty}^{\times} B(\frac{d}{dt})^j [f] \in L_2^m[\mathbb{R}^+]\}.$$

Here $N^{\times}$ is the pole at infinity of $(\lambda G - A^{\times})^{-1}$. We claim that both D_1 and D_2 are dense subspaces of $L_2^m[\mathbb{R}^+]$. Indeed, all functions $q_+(f)$ with $f \in L_2^m(\mathbb{R})$ and supp f compact belong to D_1. In fact, for such a function $q_+(f)$ one may take for x the vector

$$x = i\int_0^{\infty} e^{is\Omega^{\times}} P_0^{\times} Bf(s)ds.$$

As the set of functions with compact support is dense in $L_2^m(\mathbb{R})$, it follows that D_1 is dense in

$L_2^m[\mathbb{R}^+]$. As for the space D_2, note that D_2 contains the subspace $(\mathscr{S})^m[\mathbb{R}^+]$ of restrictions of functions in $\mathscr{S}^m$ to $(0,\infty)$. Since $\mathscr{S}^m$ is a dense subspace of $L_2^m(\mathbb{R})$, the subspace D_2 is dense in $L_2^m[\mathbb{R}^+]$, as well.

Next, we prove that $\text{Im } T_W \subset D_1$. Take an element $[f] \in \text{Im } T_W$. Let f be any representative of $[f]$ in $L_2^m(\mathbb{R})$, and let $\phi \in D(T_W)$ be such that $T_W(\phi) = [f]$. Then, by Theorem 6.1, there exists a distribution $\rho \in (\mathscr{S}^n)'$ satisfying the equations in (7.17). By a remark just after the proof of Theorem 6.1, we know that $\rho \in H_\mu^n(\mathbb{R})$ for some integer μ. Now, repeat the argument in the second paragraph of Part (A) in the proof of Lemma 7.2 and observe that $[\rho] \in H_\mu^n[\mathbb{R}^+]$ satisfies the equations

$$i\frac{d}{dt}G[\rho] = A^\times[\rho] + B[f], \quad [\phi] = -C[\rho] + [f].$$

Hence, Theorem 4.1 states that $[\rho]$ is of the form

$$\begin{aligned}[\rho] = q_+(p^\times(i\frac{d}{dt})Bf + \int_0^\infty l^\times(t-s)Bf(s)ds) + [E^\times e^{-it\Omega^\times} x_-^\times] +\\ + [E^\times e^{-it\Omega^\times}(x_0^\times - i\int_0^t e^{is\Omega^\times} P_0^\times Bf(s)ds)],\end{aligned} \tag{7.32}$$

where $p^\times$ is some $n \times n$ matrix polynomial, $l^\times$ is an $n \times n$ matrix function with components in $L_1(\mathbb{R})$ and $x_-^\times$ and $x_0^\times$ are vectors in $\text{Im } P_-^\times$ and $\text{Im } P_0^\times$, respectively. The terms in the first line of formula (7.32) are elements of the space $H_\nu^n[\mathbb{R}^+]$ for ν sufficiently small. As the same holds for $[\rho]$ itself, we must have $[\chi(Bf)] \in H_\nu^n[\mathbb{R}^+]$, where

$$\chi(Bf) := \left(e^{-it\Omega^\times}(x_0^\times - i\int_0^t e^{is\Omega^\times} P_0^\times Bf(s)ds)\right)_+.$$

We claim that, in fact, $[\chi(Bf)] \in L_2^n[\mathbb{R}^+]$. Indeed, for $\nu \geq 0$ this is clear. For negative ν employ the identity

$$i\frac{d}{dt}[\chi(Bf)] - i[\chi(Bf)] = (\Omega^\times - i)[\chi(Bf)] + [P_0^\times Bf]. \tag{7.33}$$

As $[P_0^\times Bf] \in L_2^n[\mathbb{R}^+] \subset H_\nu^n[\mathbb{R}^+]$, equation (7.33) implies that, together with $[\chi(Bf)]$, the distribution $(i\frac{d}{dt} - i)[\chi(Bf)] \in H_\nu^n[\mathbb{R}^+]$. Now, introduce an operator $\tilde{D}_- : (\mathscr{S}^n)' \to (\mathscr{S}^n)'$ by $\tilde{D}_-(f) = i\frac{d}{dt}f - if$. It is easy to check that the operator $\tilde{D}_-$ is invertible and maps $H_{\nu+1}^n(\mathbb{R})$ onto $H_\nu^n(\mathbb{R})$. A closer look reveals that $\tilde{D}_-$ maps $H_{\nu+1}^n(\overline{\mathbb{R}^-})$ onto $H_\nu^n(\overline{\mathbb{R}^-})$. Here $H_\mu^n(\overline{\mathbb{R}^-})$ denotes the space of distributions in $H_{\mu+1}^n(\mathbb{R})$ with support in $(-\infty,0]$. As a consequence, the operator $D_- : H_{\nu+1}^n[\mathbb{R}^+] \to H_\nu^n[\mathbb{R}^+]$, given by $[\rho] \to i\frac{d}{dt}[\rho] - i[\rho]$, is invertible, and hence,

$[\chi(Bf)] \in H^n_{\nu+1}[\mathbb{R}^+]$. Repeating this argument $-\nu-1$ times, we obtain that $[\chi(Bf)] \in L^n_2[\mathbb{R}^+]$, which proves the claim. But then, we have $\chi(Bf) \in L^n_2(\overline{\mathbb{R}^+})$, and so $[f] \in D_1$.

From the results in the preceding paragraph and Proposition 7.4, it follows that $D_1 = L^m_2[\mathbb{R}^+]$. Using this equality, we now prove that $P^\times_0 B = 0$. Indeed, suppose that $P^\times_0 B \neq 0$. Then there exist a number $a \in \mathbb{R}$ and vectors $0 = v_{-1}, v_0, v_1, \ldots, v_{l-1}$ such that

$$v_j P^\times_0 = v_j, \quad v_j(\Omega^\times - a) = v_{j-1} \quad (j = 0, 1, \ldots, l-1),$$

$$v_j B = 0 \quad (j = 0, 1, \ldots, l-2), \quad v_{l-1} B \neq 0.$$

Now, let $1 \leq j \leq m$ be an integer such that $c = (v_{l-1}B)_j \neq 0$, and let f be the $\mathbb{C}^m$-valued function with all components zero, except for f_j, which is given by

$$f_j(t) = e^{-ita}\frac{1}{1+t}, \quad t > 0, \qquad f_j(t) = 0, \quad t < 0.$$

The function $[f] \in L^m_2[\mathbb{R}^+] = D_1$, and hence there exists a vector $x^\times_0 \in \operatorname{Im} P^\times_0$ such that $(\chi(Bf))_+ \in L^n_2(\overline{\mathbb{R}^+})$. However, a premultiplication of the expression $(\chi(Bf))_+$ by v_{l-1} yields

$$e^{-ita}\Big(\big(v_{l-1} + (-it)v_{l-2} + \cdots + \frac{(-it)^{l-1}}{(l-1)!}v_0\big)x^\times_0 - i\int_0^t e^{isa}(v_{l-1}B)f(s)ds\Big)_+ =$$

$$= e^{-ita}\Big(\big(v_{l-1} + (-it)v_{l-2} + \cdots + \frac{(-it)^{l-1}}{(l-1)!}v_0\big)x^\times_0 - ic\log(1+t)\Big)_+ \notin L_2(\overline{\mathbb{R}^+}).$$

Thus, we arrive at a contradiction, proving that $P^\times_0 B = 0$.

To establish the $\mathbb{R}$-regularity of $\lambda G - A^\times$, or, equivalently condition (i'), it now suffices to recall formula (5.6), which states that $\Theta^\times = (\lambda G - A^\times, B, -C)$ is a realization of $W^{-1}(.)$. Hence, we have

$$W(\lambda)^{-1} = I - C(\lambda G - A^\times)^{-1}(P^\times_\infty + P^\times_- + P^\times_+)B,$$

and it follows that the function $W^{-1}(.)$ is analytic at all points $\lambda \in \mathbb{R}$, indeed.

It remains to prove (ii). Again, we first show that $D_2 = L^m_2[\mathbb{R}^+]$. To verify this, let $[f] \in \operatorname{Im} T_W$, and let f be any representative of $[f]$ in $L^m_2(\mathbb{R})$. Note that $P^\times_0 = 0$ because of the $\mathbb{R}$-regularity of $\lambda G - A^\times$. Hence, employing (7.17) and (7.32), there exists a vector $x^\times_-$ in the image of $P^\times_-$ such that

$$q_+(Cp^\times(i\frac{d}{dt})Bf + \int_0^\infty C\Gamma^\times(t-s)Bf(s)ds) + [CE^\times e^{-it\Omega^\times}x^\times_-] \in L^m_2[\mathbb{R}^+].$$

As the last two terms in the above formula are functions in $L^m_2[\mathbb{R}^+]$, the same has to be true for $Cp^\times(i\frac{d}{dt})B[f]$. But then, since

$$p^{\times}(\lambda)=-\sum_{j=0}^{N^{\times}} E^{\times}(\Omega^{\times})^{j} P_{\infty}^{\times} B\lambda^{j},$$

the function $[f]\in D_2$. Thus Im $T_W \subset D_2$, and hence by Proposition 7.4 $D_2 = L_2^m[\mathbb{R}^+]$. Next, take an arbitrary vector $x\in \mathbb{C}^m$, and define a function f by $f(t)=x$ for $0<t<1$ and $f(t)=0$ for other values of t. Then, because $[f]\in D_2$,

$$q_+(CE^{\times}P_{\infty}^{\times}Bf)-q_+(\sum_{j=1}^{N^{\times}} CE^{\times}(i\Omega^{\times})^{j}P_{\infty}^{\times}B(\frac{d}{dt})^{j-1}\delta_1 x)\in L_2[\mathbb{R}^+]. \tag{7.34}$$

Here δ_1 is the distribution $\delta_1 \in \mathcal{S}'$, given by $\langle\delta_1,\alpha\rangle=\alpha(1)$ for $\alpha\in\mathcal{S}$. Since the first term in (7.34) is an element of $L_2^m[\mathbb{R}^+]$, it follows that

$$\sum_{j=1}^{N^{\times}} CE^{\times}(i\Omega^{\times})^{j}P_{\infty}^{\times}B(\frac{d}{dt})^{j-1}\delta_1 x+T_- = g\in L_2^m(\mathbb{R})$$

for some $T_- \in (\mathcal{S}^m)'$ with support in $(-\infty,0]$. Now, decompose the function g as $g=g_+ + g_-$, where $g_\pm \in L_2^m(\mathbb{R})$ and $g_\pm$ vanishes for $\pm t<0$. Thus, one obtains

$$\sum_{j=1}^{N^{\times}} CE^{\times}(i\Omega^{\times})^{j}P_{\infty}^{\times}B(\frac{d}{dt})^{j-1}\delta_1 x-g_+ = g_- - T_- \in (\mathcal{S}^m)'(\overline{\mathbb{R}^+})\cap(\mathcal{S}^m)'(\overline{\mathbb{R}^-}).$$

In particular, there exists a tempered distribution T_0 with support in $\{0\}$ such that

$$\sum_{j=1}^{N^{\times}} CE^{\times}(i\Omega^{\times})^{j}P_{\infty}^{\times}B(\frac{d}{dt})^{j-1}\delta_1 x+T_0\in L_2^m(\overline{\mathbb{R}^+})\subset L_2^m(\mathbb{R}).$$

Applying Fourier transformation to the above expression yields

$$e^{i\lambda}\sum_{j=1}^{N^{\times}} iCE^{\times}(\Omega^{\times})^{j}P_{\infty}^{\times}Bx\lambda^{j-1}+\mathcal{F}(T_0)\in L_2^m(\mathbb{R}),$$

where $\mathcal{F}(T_0)$ is a polynomial (see [Tr], Theorem 24.6). But then, a straightforward reasoning implies that $CE^{\times}(\Omega^{\times})^{j}P_{\infty}^{\times}Bx=0$ for $j\geq 1$. Since x is an arbitrary vector in $\mathbb{C}^m$, this proves condition (ii).

Step (iii). Assume the conditions (i) and (ii) hold. By Lemma 7.1, the kernel of T_W then is given by (1.16) and has a finite dimension given by (1.17). Furthermore, Lemma 7.2 states that the image of T_W is the closed subspace given in (1.18) with finite codimension as in (1.19). Since T_W is a closed operator, this proves that T_W is Fredholm. To compute its index, note that

$$\text{codim Im } T_W = \dim(\text{Im } P_-^{\times}\cap(\text{Ker } P_- + \text{Im } \mathcal{B}))+n-\dim(\text{Ker } P_- + \text{Im } \mathcal{B})-\dim \text{Im } P_-^{\times}.$$

Thus, the index is given by

$$\text{ind } T_W = \dim \text{Im } P_-^{\times} - n + \dim(\text{Ker } P_- + \text{Im } \mathcal{B}) + \dim \text{Ker } \mathcal{B} +$$

$$+\dim(\operatorname{Ker} P_- \cap \operatorname{Im} \mathscr{B}) = \dim \operatorname{Im} P_-^\times - \dim \operatorname{Im} P_- - ms,$$

which proves (1.20). It remains to prove that (1.21) satisfies (1.22), or in other words to prove that (1.21) is a solution to the equation $T_W(\phi) = q_+(f)$, with f a function in $H^m_{s+r}(\mathbb{R})$ satisfying the condition in (1.18). Define the vectors x and v_j, $j=0,1,..,-s-1$, as in Theorem 1.3. Then $x \in \operatorname{Im} P_-^\times$. Furthermore, as

$$\operatorname{Im} \Lambda_2^+ \subset \operatorname{Im} \mathscr{B},$$

the definition of $\mathscr{B}^+$ implies that

$$i\sum_{j=0}^{-s-1} \Omega^j B v_j = -\Lambda_2^+ P_- \Lambda_1^+ (I - P_-^\times) R_{r,s}(Bf).$$

Therefore, by the definitions of Λ_2^+ and Λ_1^+, we have

$$P_-(i\sum_{j=0}^{-s-1} \Omega^j B v_j) = -P_- \Lambda_1^+ (I - P_-^\times) R_{r,s}(Bf). \tag{7.35}$$

Next, employ (1.18) to conclude that $(I - P_-^\times) R_{r,s}(Bf) = (I - P_-^\times) u$ for some vector u in the space $\operatorname{Ker} P_- + \operatorname{Im}(B\,\Omega B \ldots \Omega^{-s-1} B)$. Hence,

$$\begin{aligned}\Lambda_1^+ (I - P_-^\times) R_{r,s}(Bf) &= (I - P_-^\times + P_-^\times) \Lambda_1^+ (I - P_-^\times) u = \\ &= (I - P_-^\times) R_{r,s}(Bf) + P_-^\times \Lambda_1^+ (I - P_-^\times) R_{r,s}(Bf).\end{aligned} \tag{7.36}$$

Combining (7.35) and (7.36) we arrive at

$$R_{r,s}(Bf) + x + i\sum_{j=0}^{-s-1} \Omega^j B v_j = (I - P_-^\times) R_{r,s}(Bf) +$$

$$+ P_-^\times \Lambda_1^+ (I - P_-^\times) R_{r,s}(Bf) + (I - P_- + P_-)(i\sum_{j=0}^{-s-1} \Omega^j B v_j) =$$

$$(I - P_-)\Lambda_1^+ (I - P_-^\times) R_{r,s}(Bf) + (I - P_-)(i\sum_{j=0}^{-s-1} B\Omega^j v_j) \in \operatorname{Ker} P_-.$$

An application of Lemma 7.2 now finishes the proof. □

Under the condition that $\lambda G - A^\times$ is $\mathbb{R}$-regular, the space D_2 introduced in Step (ii) of the above proof can be identified with the domain of the maximal integro-differential operator $T_{W^{-1}}$ $(L_2^m(\overline{\mathbb{R}^+}) \to L_2^m[\mathbb{R}^+])$. Using this observation it is possible to give an alternative proof of the inclusion $\operatorname{Im} T_W \subset D_2$, and hence, proceeding as in Step (ii), of the necessity of condition (ii) for T_W to be Fredholm.

PROOF OF THEOREM 1.2. If the operator T_W is invertible, then the operator T_W is Fredholm, injective and surjective. Thus, by Theorem 1.3, the conditions (i) and (ii) hold, and the right hand sides in (1.17) and (1.19) are zero. In particular, the first two terms in (1.17) are zero, which together with (1.19) implies condition (iv). Finally, the matrix $\mathcal{B}$ is injective because the third term in (1.17) is zero. This implies condition (iii). Conversely, if the conditions (i) up to (iv) hold, then the operator T_W is Fredholm and, using the formulas in Theorem 1.3, we have

$$\text{codim Im } T_W = 0 = \dim \text{Ker } T_W .$$

So, T_W is invertible. To prove (1.14), it suffices to check that the vectors x and v_j, $j=0,1,..,-s-1$, defined in (1.15), satisfy the condition in (7.14). Indeed, since Im $\Pi_2 \subset$ Im $\mathcal{B}$, we have $\Pi_2 R_{r,s}(Bf) = \mathcal{B}\mathbf{u}$ for some $\mathbf{u} \in \mathbb{C}^{-ms}$. Hence, it follows that

$$R_{r,s}(Bf) + x + i \sum_{j=0}^{-s-1} \Omega^j B v_j = R_{r,s}(Bf) + x - \mathcal{B}\mathcal{B}^{(-1)}\mathcal{B}\mathbf{u} =$$

$$= R_{r,s}(Bf) + x - \mathcal{B}\mathbf{u} = (I - \Pi_1 - \Pi_2) R_{r,s}(Bf) \in \text{Ker } P_- .$$

This finishes the proof. □

8. AN EXAMPLE

Let us illustrate the results of the previous sections by considering the system of half-line equations

$$\begin{cases} q_+(\tfrac{1}{2}k * \phi_1 + \phi_2) = [f_1], \\ q_+(ik_- * \phi_1 + i\frac{d}{dt}\phi_2 - i\phi_2) = [f_2]. \end{cases} \tag{8.1}$$

Here $k(t) = e^{-|t|}$ and $k_-(t) = k(t) - k_+(t)$. The equations (8.1) have as their symbol the function

$$W(\lambda) = \begin{bmatrix} (\lambda^2+1)^{-1} & 1 \\ (\lambda - i)^{-1} & \lambda - i \end{bmatrix}.$$

The determinant of the function W equals

$$\det W(\lambda) = -2i(\lambda^2+1)^{-1}.$$

In particular, $W(\lambda)$ is invertible for all real λ, and a straightforward computation shows that $W(\lambda)^{-1}$ is the polynomial

$$W(\lambda)^{-1} = \tfrac{1}{2}i \begin{bmatrix} (\lambda^2+1)(\lambda - i) & -(\lambda^2+1) \\ -(\lambda+i) & 1 \end{bmatrix}.$$

So, it follows, *e.g.* by taking the Fourier transformation, that the operator $\tilde{L}_W : (\mathcal{S}^2)' \to (\mathcal{S}^2)'$ is invertible. Also, we see that the conditions (i') and (ii') are fulfilled for $r \geq 3$, which implies that the maximal half-line operator T_W $(H_s^2(\mathbb{R}^+) \to H_{s+r}^2[\mathbb{R}^+])$ with symbol W is Fredholm for $r \geq 3$. We shall compute the Fredholm characteristics of the operator T_W for $r = 3$ and $s = 0, -1, -2$ and -3, respectively, and give formulas for possible generalized inverses.

To analyse the Fredholm properties of the operator T_W, we first write W in realized form, *i.e.*, in the form (1.9). Put

$$G = \begin{bmatrix} 0&1&0&0&0 \\ 0&0&0&0&0 \\ 0&0&0&0&0 \\ 0&0&0&1&0 \\ 0&0&0&0&1 \end{bmatrix}, \quad A = \begin{bmatrix} 1&0&0&0&0 \\ 0&1&0&0&0 \\ 0&0&1&0&0 \\ 0&0&0&i&0 \\ 0&0&0&0&-i \end{bmatrix}, \quad B = \begin{bmatrix} 0 & 1+i \\ 0 & -1 \\ 1 & -1 \\ 1 & 0 \\ 1 & 0 \end{bmatrix}, \quad C = \begin{bmatrix} 0 & 0 & 1 & -\tfrac{1}{2}i & \tfrac{1}{2}i \\ 1 & 0 & 0 & 1 & 0 \end{bmatrix}.$$

Then a straightforward calculation verifies that, for the above matrices, the identity (1.9) holds.

The next step is to compute the data for the decompositions of the pencils $\lambda G - A$ and $\lambda G - A^\times$. As for the pencil $\lambda G - A$, Theorem 1.2 shows that we need only the projection P_- and the matrix Ω. These matrices are given by

$$P_- = \begin{bmatrix} 0&0&0&0&0 \\ 0&0&0&0&0 \\ 0&0&0&0&0 \\ 0&0&0&0&0 \\ 0&0&0&0&1 \end{bmatrix}, \quad \Omega = \begin{bmatrix} 0&1&0&0&0 \\ 0&0&0&0&0 \\ 0&0&0&0&0 \\ 0&0&0&i&0 \\ 0&0&0&0&-i \end{bmatrix}.$$

In particular, we have

$$\operatorname{Ker} P_- = \operatorname{span} \{e_1, e_2, e_3, e_4\},$$

where e_j is the jth unit vector of $\mathbb{C}^5$. Next, we direct our attention to the pencil $\lambda G - A^\times$, where

$$A^\times = A - BC = \begin{bmatrix} -i & 0 & 0 & -1-i & 0 \\ 1 & 1 & 0 & 1 & 0 \\ 1 & 0 & 0 & \tfrac{1}{2}i+1 & -\tfrac{1}{2}i \\ 0 & 0 & -1 & 1\tfrac{1}{2}i & -\tfrac{1}{2}i \\ 0 & 0 & -1 & \tfrac{1}{2}i & -1\tfrac{1}{2}i \end{bmatrix}.$$

The pencil $\lambda G - A^\times$ has determinant $-2i$, and hence is regular and has no zeros in the complex plane. Therefore, we immediately have

$$P_-^\times = P_+^\times = P_0^\times = P^\times = 0, \quad P_\infty^\times = I_5.$$

To compute the matrices $E^\times$ and $\Omega^\times$, we first invert the matrix $\lambda G - A^\times$. To do so, we use the formula manipulation program Maple, which yields

$$(\lambda G - A^{\times})^{-1} = \frac{1}{2\pi i} \begin{bmatrix} -(\lambda+i-1) & -(\lambda^2+(i-1)\lambda) & \lambda^2-\lambda+1-i & \tfrac12 i\lambda-\tfrac12 i+\tfrac12 & -\tfrac12\lambda+\tfrac12 i-\tfrac12 \\ -1 & -\lambda-2i & \lambda+i & \tfrac12 i & -\tfrac12 i \\ -\lambda^2-2 & -\lambda^3-2\lambda & \lambda^3-i\lambda^2+2\lambda-2i & \tfrac12 i\lambda^2+\lambda+2i & -\tfrac12 i\lambda^2-\lambda \\ \lambda+i & \lambda^2+i\lambda & -\lambda^2-1 & -\tfrac12 i\lambda-\tfrac12 & \tfrac12 i\lambda+\tfrac12 \\ \lambda-i & \lambda^2-i\lambda & -\lambda^2+2i\lambda+1 & -\tfrac12 i\lambda-1\tfrac12 & \tfrac12 i\lambda+1\tfrac12 \end{bmatrix}.$$

Using this formula, a straightforward calculation involving contour integration shows that

$$E^{\times} = \begin{bmatrix} \tfrac12 i+\tfrac12 & 0 & \tfrac12 i+\tfrac12 & \tfrac14 i+\tfrac14 & -\tfrac14-\tfrac14 i \\ -\tfrac12 i & 1 & -\tfrac12 & -\tfrac14 & \tfrac14 \\ -i & 0 & 1 & -1 & 0 \\ -\tfrac12 & 0 & -\tfrac12 i & -\tfrac14 i & \tfrac14 i \\ \tfrac12 & 0 & \tfrac12 i & -\tfrac34 i & \tfrac34 i \end{bmatrix}, \quad \Omega^{\times} = \begin{bmatrix} -\tfrac12 i & 1 & -\tfrac12 & -\tfrac14 & \tfrac14 \\ 0 & 0 & 0 & 0 & 0 \\ 0 & 0 & 0 & 0 & 0 \\ -\tfrac12 & 0 & -\tfrac12 i & -\tfrac14 i & \tfrac14 i \\ \tfrac12 & 0 & \tfrac12 i & \tfrac34 i & -\tfrac34 i \end{bmatrix}.$$

We are now in a position to compute kernel and image of T_W. Firstly, verify the identities

$$P_-^{\times} = 0, \quad \text{codim Ker } P_- = 1, \quad \text{Ker } P_- + \text{Im } B = \mathbb{C}^5.$$

Inserting these formulas into Theorem 1.3, the Fredholm indices of T_W are given by the following table:

s	dim Ker T_W	codim Im T_W	ind T_W
0	0	1	−1
−1	1	0	1
−2	3	0	3
−3	5	0	5

In particular, for $s=0$ the operator T_W is injective. As for the image, note that $P_+^{\times}=0$. Hence, by the formula in Theorem 1.3, the image Im T_W equals

$$\{q_+(f) \mid f \in H_3^2(\mathbb{R}), -\tfrac12 if_1(0) - \tfrac12 f_2(0) + if_1'(0) + \tfrac12 f_2'(0) - \tfrac12 f_1^{(2)}(0) = 0\}.$$

Here $f = (f_1 \; f_2)^T$. For $s<0$ the mapping T_W is surjective, and it remains to compute the kernel. Observe that the vector x in Theorem 1.3 must be the zero vector because $P_-^{\times}=0$. As for the vectors v_j, the condition that $\sum_{j=0}^{-s-1} \Omega^j B v_j \in \text{Ker } P_-$ comes down to requiring the last component in the summation to be zero. Thus, an easy calculation leads to

$$\text{Ker } T_W = \{ \begin{pmatrix} v_1 \\ v_2 \end{pmatrix} \delta \mid v_1 = 0 \} = \{ \begin{pmatrix} 0 \\ \alpha \end{pmatrix} \delta \mid \alpha \in \mathbb{C} \}$$

in case $s=-1$,

$$\operatorname{Ker} T_W = \{ \begin{pmatrix} v_1 \\ v_2 \end{pmatrix} \delta + i \begin{pmatrix} w_1 \\ w_2 \end{pmatrix} \frac{d}{dt}\delta \mid v_1 - iw_1 = 0 \}$$

in case $s=-2$, and

$$\operatorname{Ker} T_W = \{ \begin{pmatrix} v_1 \\ v_2 \end{pmatrix} \delta + i \begin{pmatrix} w_1 \\ w_2 \end{pmatrix} \frac{d}{dt}\delta - \begin{pmatrix} z_1 \\ z_2 \end{pmatrix} (\frac{d}{dt})^2 \delta \mid v_1 - iw_1 - z_1 = 0 \}$$

in case $s=-3$.

Finally, let us compute formulas for a possible generalized inverse of T_W. To deal with the case $s=0$, recall that $P_-^\times = P_+^\times = 0$. Hence, by Theorem 1.3 it remains to verify that

$$I + CE^\times P_\infty^\times B = \begin{bmatrix} ½ & -½i \\ ½ & ½i \end{bmatrix}, \quad CE^\times \Omega^\times P_\infty^\times B = \begin{bmatrix} ½i & 0 \\ -½i & 0 \end{bmatrix},$$

$$CE^\times (\Omega^\times)^2 P_\infty^\times B = \begin{bmatrix} ½ & -½i \\ 0 & 0 \end{bmatrix}, \quad CE^\times (\Omega^\times)^3 P_\infty^\times B = \begin{bmatrix} ½i & 0 \\ 0 & 0 \end{bmatrix}.$$

Observe that the above four matrices are just the coefficients of λ^j for $j=0,1,2,3$ of the polynomial $W(\lambda)^{-1}$. From the above results one immediately obtains the following formula for a generalized inverse of T_W for $s=0$:

$$T_W^+(q_+(f)) = \begin{bmatrix} ½f_1 - ½if_2 - ½f_1' - ½f_1^{(2)} + ½if_2^{(2)} + ½f_1^{(3)} \\ ½f_1 + ½if_2 + ½f_1' \end{bmatrix}_+ ,$$

where $f = (f_1 \; f_2)^T \in H_3^2(\mathbb{R})$. Note that, in this case, the operator T_W^+ is a left inverse of T_W. To compute a generalized inverse in case $s<0$ we need the operators Λ_1^+, Λ_2^+, $\mathcal{B}^+$ and $R_{3,s}$. Recall that $\operatorname{Ker} P_- + \operatorname{Im} B = \mathbb{C}^5$, and that $P_-^\times = 0$. Hence, for any negative s one may take the identity matrix of $\mathbb{C}^5$ for Λ_1^+. For Λ_2^+ one may choose, independent of the parameter s, the operator mapping the vector $\alpha.e_5$ in $\operatorname{Im} P_-$ to the vector $(0,0,\alpha,\alpha,\alpha)^T$ in the image of B. It remains to find $\mathcal{B}^+$ and $R_{3,s}$. For $s=-1$, the operator $\mathcal{B}^+$ may be chosen as a left inverse of B, *e.g.*

$$\mathcal{B}^+ = \begin{bmatrix} 0 & 0 & 0 & 0 & 1 \\ 0 & -1 & 0 & 0 & 0 \end{bmatrix}.$$

Furthermore, a straightforward calculation shows that

$$R_{3,-1}(Bf) = \begin{bmatrix} * & * & * & * & -if_1(0) - ½f_2(0) + ½if_1'(0) \end{bmatrix}^T,$$

with $*$ an irrelevant element. Thus one derives the following generalized inverse of $T_W\,(H_{-1}^2(\mathbb{R}^+) \to H_2^2[\mathbb{R}^+])$:

$$T_W^+(q_+(f)) = \begin{bmatrix} ½f_1 - ½if_2 - ½f_1' - ½f_1^{(2)} + ½if_2^{(2)} \\ ½f_1 + ½if_2 + ½f_1' \end{bmatrix}_+ +$$

$$+\begin{bmatrix} \tfrac12\frac{d}{dt}((f_1^{(2)})_+) \\ 0 \end{bmatrix} + \begin{bmatrix} \tfrac12 f_1(0) - \tfrac12 i f_2(0) - f_1'(0) + \tfrac12 i f_2'(0) \\ \tfrac12 f_1(0) \end{bmatrix}\delta,$$

with $f=(f_1\ \ f_2)^T \in H_{.}^2(\mathbb{R})$. In case $s=-2$, the matrix $\mathscr{B}^+$ may be chosen as a left inverse of $(B\ \ \Omega B)$, say

$$\mathscr{B}^+ = \begin{bmatrix} 0 & 0 & 0 & \tfrac12 & \tfrac12 \\ 0 & -1 & 0 & 0 & 0 \\ 0 & 0 & 0 & -\tfrac12 i & \tfrac12 i \\ -1 & -1-i & 0 & 0 & 0 \end{bmatrix}.$$

The vector $R_{3,-2}(Bf)$ has as its last component $-\tfrac12 f_1(0)$. Thus, a generalized inverse of $T_W\ (H_{-2}^2(\mathbb{R}^+) \rightarrow H_1^2[\mathbb{R}^+])$ is given by

$$T_W^+(q_+(f)) = \begin{bmatrix} \frac{d}{dt}((-\tfrac12 f_1' + \tfrac12 i f_2')_+) + \tfrac12(\frac{d}{dt})^2((f_1')_+) \\ 0 \end{bmatrix} + \begin{bmatrix} 0 \\ \tfrac12 f_1(0) \end{bmatrix}\delta +$$

$$+\begin{bmatrix} \tfrac12 f_1 - \tfrac12 i f_2 - \tfrac12 f_1' \\ \tfrac12 f_1 + \tfrac12 i f_2 + \tfrac12 f_1' \end{bmatrix}_+ + \begin{bmatrix} -\tfrac12 f_1(0) + \tfrac12 i f_2(0) \\ 0 \end{bmatrix}\frac{d}{dt}\delta.$$

Here $f=(f_1\ \ f_2)^T \in H_1^2(\mathbb{R})$. To compute a generalized inverse for T_W when $s=-3$, observe that $(\Omega^\times)^4=0$, and hence $R_{3,-3}=0$. Thus, the vectors v_j, $j=0,1,2$ are all zero, and as generalized inverse of T_W in case $s=-3$ we find

$$T_W^+(q_+(f)) = \begin{bmatrix} \tfrac12 f_1 - \tfrac12 i f_2 \\ \tfrac12 f_1 + \tfrac12 i f_2 \end{bmatrix}_+ + \begin{bmatrix} -\tfrac12\frac{d}{dt}((f_1)_+) + \tfrac12(\frac{d}{dt})^2((f_2-f_1)_+) + \tfrac12(\frac{d}{dt})^3((f_1)_+) \\ \tfrac12\frac{d}{dt}((f_1)_+) \end{bmatrix},$$

where $f=(f_1\ \ f_2)^T \in L_2^2(\mathbb{R})$. In all three cases $s=-1,-2,-3$, the formulas for T_W^+ define a right inverse for the operator T_W.

REFERENCES

[BGK1] Bart, H., Gohberg, I. and Kaashoek, M. A.: Wiener-Hopf integral equations, Toeplitz matrices and linear systems, in "Toeplitz Centennial" (Ed. I. Gohberg), Operator Theory: Advances and Applications, Vol. 4, Birkhäuser Verlag, Basel, 1982, pp. 85-135.

[BGK2] Bart, H., Gohberg, I. and Kaashoek, M. A.: Convolution equations and linear systems, *Integral Equations and Operator Theory* **5** (1982), 283-340.

[BGK3] Bart, H., Gohberg, I. and Kaashoek, M. A.: Explicit Wiener-Hopf factorization and realization, in "Constructive methods for Wiener-Hopf factorization" (Eds. I. Gohberg and M. A. Kaashoek), Operator Theory: Advances and Applications, Vol. 21, Birkhäuser Verlag, Basel, 1986, pp. 235-316.

[BGK4] Bart, H., Gohberg, I. and Kaashoek, M. A.: The state space method in problems of analysis, *Proc. ICIAM 87*, Paris-La Villette, 1987.

[GGK] Gohberg, I., Goldberg, S. and Kaashoek, M. A.: Classes of Linear Operators Vol. 1, Operator Theory: Advances and Applications, Vol. 49, Birkhäuser Verlag, Basel 1990.

[GK1] Gohberg, I. and Kaashoek, M. A.: Block Toeplitz operators with rational symbols, in "Contributions in Operator Theory and its Applications" (Eds. I. Gohberg, J.W. Helton, L. Rodman), Operator Theory: Advances and Applications, Vol. 35, Birkhäuser Verlag, Basel, 1988, pp. 385-440.

[GK2] Gohberg, I. and Kaashoek, M. A.: The state space method for solving singular integral equations, in "Mathematical system theory, the influence of R. E. Kalman" (Ed. A. C. Antoulas), Springer Verlag, Berlin, etc., 1991.

[GKLR] Gohberg, I., Kaashoek, M. A., Lerer, L. and Rodman, L.: Common multiples and common divisors of matrix polynomials I: spectral method, *Ind. J. Math.* **30**, no. 3 (1981), 321-356.

[GKr] Gohberg, I. and Kreĭn, M. G.: Systems of integral equations on a half line with kernels depending on the difference of arguments, *Uspehi Mat. Nauk* **13** (1958), no. 2 (80), 3-72, *(Russian)* = *Amer. Math. Soc. Transl.* (2) **14** (1960), 217-287.

[GLR] Gohberg, I., Lancaster, P. and Rodman, L.: Matrix polynomials, Academic Press, New York, etc., 1982.

[H] Hörmander, L.: Linear partial differential operators, Springer Verlag, Berlin, etc., 1963.

[Ka] Kailath, T.: Linear systems, Prentice-Hall, Englewood Cliffs, NJ, 1980.

[Kr] Kreĭn, M. G.: Integral equations on a half-line with kernel depending upon the difference of the arguments, *Uspehi Mat. Nauk* **13** (1958), no. 5 (83), 3-120, *(Russian)* = *Amer. Math. Soc. Transl.* (2) **22** (1962), 163-288.

[Ku] Kuijper, A. B.: The state space method for integro-differential equations of convolution type with rational matrix symbols, Ph. D. thesis, Vrije Universiteit Amsterdam, Amsterdam, 1992.

[R1] Ramm, A. G.: Theory and Applications of some new classes of integral equations, Springer Verlag, New York, etc., 1980.

[R2] Ramm, A. G.: Random fields estimation theory, *Mathl. Comput. Modelling* **13**, No. 4 (1990), 87-100.

[Ro] Roozemond, L.: Systems of non-normal and Wiener-Hopf equations, Ph. D. thesis, Vrije Universiteit Amsterdam, Amsterdam, 1987.

[S] Schwartz, L.: Théorie des distributions, Hermann, Paris, 1966.

[St] Stummel, F.: Diskrete Konvergenz linearer Operatoren II, *Math. Z.* **120** (1971), 231-264.

[Ta] Talenti, G.: Sulle equazioni integrali di Wiener-Hopf, *Boll. Un. Mat. Ital.* (4) **7**, suppl. fasc. 1 (1973), 18-118.

[Tr] Treves, F.: Topological vector spaces, distributions and kernels, Academic Press, New York, 1967.

[VG1] Volevich, L. R., and Gindikin, S. G.: The Cauchy problem and other related problems for convolution equations, *Uspehi Mat. Nauk* **27** (1972), no. 4 (166), 65-143 *(Russian)* = *Russian Math. Surveys* **27** (1972), 71-160.

[VG2] Volevich, L. R., and Gindikin, S. G.: The Wiener-Hopf equation in generalized functions, *Trudy Moskov. Mat. Obšč.* **35** (1976), 165-214 *(Russian)* = *Trans. Moscow Math. Soc.* (1979), no. 1 (**35**), 167-216.

[VG3] Volevich, L. R., and Gindikin, S. G.: The Wiener-Hopf equation in generalized functions that are smooth on the half line, *Trudy Moskov. Mat. Obšč.* **38** (1979), 29-74 *(Russian)* = *Trans. Moscow Math. Soc.* (1980), no. 2 (**38**), 25-70.

[VG4] Volevich, L. R., and Gindikin, S. G.: The Wiener-Hopf equation for smooth and generalized functions in a half-space, *Trudy Moskov. Mat. Obšč.* **40** (1979), 241-295 *(Russian)* = *Trans. Moscow Math. Soc.* (1981), no. 2 (**40**), 231-284.

[VG5] Volevich, L. R., and Gindikin, S. G.: Boundary value problems for the Wiener-Hopf equation in a half-space, *Trudy Moskov. Mat. Obšč.* **42** (1981) *(Russian)* = *Trans. Moscow Math. Soc.* (1982), no. 2 (**42**), 175-198.

Department of Mathematics and Computer Science
Vrije Universiteit
de Boelelaan 1081a
1081 HV Amsterdam
The Netherlands

MSC 1991
Primary: 45J05, 45E10
Secondary: 15A22, 93C05

Operator Theory:
Advances and Applications, Vol. 58

SYMBOLS AND ASYMPTOTIC EXPANSIONS

Harold Widom[1]

A general principle is proposed that in all the usual asymptotic expansions for tr $f(T)$ where f is a "general" function and T is a Toeplitz, Wiener-Hopf, or pseudodifferential operator, each term of the expansion is an integral of (or, more generally, some distribution applied to) $f(\sigma^*)$, where σ^* is a "Symbol" associated with that term of the expansion. These Symbols are defined in terms of σ, the symbol of the given operator, but have larger domain. Although nothing so general is proved we consider several operators with smooth or nonsmooth symbol, evaluate early terms of asymptotic expansions associated with them, and show that they can all be cast into the form described.

0. INTRODUCTION

Since its earliest years the study of Toeplitz and Wiener-Hopf operators has focused on questions of invertibility and Fredholmness on the one hand and the asymptotics of finite sections on the other. One of the fundamental early results is that the Toeplitz operator $T(\varphi)$ with (scalar-valued) continuous symbol φ is Fredholm if and only if φ is nonzero on the circle $\mathbb{T}$, and is invertible if and only if the winding numbers of φ equals zero. For their study of operators with piecewise continuous symbols, Gohberg and Krupnik [GK] introduced a function φ^* which has domain $\mathbb{T} \times [0, 1]$ and is defined by

(0.1) $$\varphi^*(\theta, t) = (1-t)\,\varphi(\theta-) + t\,\varphi(\theta+), \qquad 0 \le t \le 1.$$

(As usual we identify $\theta \in \mathbb{R}$ with $e^{i\theta} \in \mathbb{T}$.) It turns out that $T(\varphi)$ is Fredholm on H_2 if and only if φ^* is nonzero, and is invertible if and only if a suitably defined winding number of φ^* is zero. This object φ^* Gohberg and Krupnik called the symbol of the operator but we prefer to call it the Symbol. Thus, the symbol φ is the function defining the operator whereas the Symbol φ^* arises in the study of its qualitative behavior.

---

1) Research supported by National Science Foundation Grant DMS 8822906.

The basic limit theorem is that, denoting by $T_n(\varphi)$ the usual $n \times n$ section of the matrix for $T(\varphi)$, one has for quite general symbols φ and functions f an asymptotic formula as $n \to \infty$

$$\text{tr}\, f(T_n(\varphi)) = \frac{n}{2\pi} \int f(\varphi(\theta))\, d\theta + o(n). \tag{0.2}$$

(If φ is real-valued then f need only be continuous; otherwise f is required to be analytic in a convex set containing the essential range of φ. In the sequel we always take our functions f to be analytic in an appropriate domain.) If φ is smooth then the term $o(n)$ in (0.2) can be sharpened to: constant + $o(1)$. We will give an expression for the constant in a moment but first we tell what happens when φ is only piecewise smooth with jump discontinuities. The sharpening of $o(n)$ in (0.2) in this case is: constant $\cdot \log n + o(\log n)$. The constant here is equal to

$$\frac{1}{4\pi^2} \sum_i U(\varphi(\theta_i -), \varphi(\theta_i +), f) \tag{0.3}$$

where the θ_i are the points of discontinuity of φ and where U is defined by

$$U(a, b, f) = \int_0^1 \frac{f((1-t)\,a + t\,b) - [(1-t) f(a) + t f(b)]}{t(1-t)}\, dt. \tag{0.4}$$

(The Wiener-Hopf version of this is in [W2]. A sharp form of the stated result, wherein φ is only required to be of bounded variation, is in [F]. Of course, the sum (0.3) might then be infinite, but still convergent.)

Basor [B] pointed out that (0.3) could be interpreted as an integral over the domain of the Gohberg-Krupnik Symbol φ^* of $T(\varphi)$. In fact this constant is the integral of $f(\varphi^*)$, adjusted by subtracting the function which is linear in t and has the same values at $t = 0$ and 1 (we call this the "linear part" of $f(\varphi^*)$), with respect to the measure on $\mathbb{T} \times [0, 1]$ which is the product of counting measure on $\mathbb{T}$ and the measure

$$\frac{1}{4\pi^2} \frac{dt}{t(1-t)} \tag{0.5}$$

on [0, 1]. (Thus we have a linear functional applied to $f(\varphi^*)$ which is a regularized integral like those familiar in distribution theory.)

We see, therefore, that although the Symbol was introduced for the study of qualitative properties of $T(\varphi)$ it arises also in the answer to a quantitative question about these operators.

If φ is smooth then of course (0.3) vanishes and the term $o(n)$ in (0.2) takes the form: constant + $o(1)$. The constant here can be written in several ways. (It was obtained first, in a different form, in [L].) We prefer, for the context of this paper,

$$\frac{1}{16\pi^2} \int\int U(\varphi(\theta), \varphi(\psi), f) \frac{d\theta\, d\psi}{\sin^2 \frac{1}{2}(\theta - \varphi)}. \tag{0.6}$$

This also can be written as a regularized integral of $f(\varphi^*)$ for an appropriately defined Symbol φ^*. Its domain is $\mathbb{T} \times \mathbb{T} \times [0, 1]$ and it is given by

(0.7) $$\varphi^*(\theta, \psi, t) = (1-t)\,\varphi(\theta) + t\,\varphi(\psi),$$

and (0.6) is the regularized integral of $f(\varphi^*)$ with respect to the product of the measure

$$\frac{1}{4}\operatorname{cosec}^2 \frac{1}{2}(\theta - \psi)\, d\theta\, d\psi$$

on $\mathbb{T} \times \mathbb{T}$ and the same measure (0.5) on [0, 1]. Why it is just these measures which occur is, as of this writing, a mystery. More such mysteries are to come.

There seems to be reason to think that the following speculation has some merit: in all the usual asymptotic expansions for tr $f(T)$, where T is a Toeplitz, Wiener-Hopf, or pseudodifferential operator, each term of the expansion is the integral of (or another distribution applied to) $f(\sigma^*)$ where σ^* is some Symbol of T associated with this term of the expansion. The bulk of this paper consists of discussion of four other examples where this is the case. We describe these now.

I. Let T be the integral operator on $L_2(\mathbb{R}^n)$, depending on the large parameter α, with kernel given by

(0.8) $$\left(\frac{\alpha}{2\pi}\right)^n \int e^{i\alpha\xi\cdot(x-y)}\sigma(x,\ \xi)\ d\xi$$

where, to simplify matters, we assume σ is a Schwartz function on $\mathbb{R}^n \times \mathbb{R}^n$. (Thus T is a pseudodifferential operator with symbol, in the sense of that theory, equal to $\sigma(x, \xi/\alpha)$.) Then there is a complete asymptotic expansion as $\alpha \to \infty$

(0.9) $$\operatorname{tr} f(T) \sim \sum_{k=0}^{\infty} c_k\, \alpha^{n-k}$$

where each c_k is given by an integral formula involving derivatives of f applied to σ, times products of derivatives of σ itself. (We require of f, in addition to analyticity in the convex hull of the range of σ, that it vanish at 0 so that $f(T)$ is trace class. This expansion was first derived in [W1], and can be found in [W4] also. In both cases the context was somewhat different: in the first the setting was a compact manifold and in the second σ vanished for x outside a compact set in $\mathbb{R}^n$ but was only required to be a symbol of negative order rather than a Schwarz function. The condition on the x-support of σ could easily have been weakened to obtain the result as stated above.) We have

(0.10) $$c_0 = \left(\frac{1}{2\pi}\right)^n \int\int f(\sigma(x,\ \xi))\ dx\ d\xi,$$

as one expects in all expansions like this. We shall show here that if we define a Symbol σ^* with domain $(\mathbb{R}^n \times \mathbb{R}^n)^2 \times [0, 1]$ by

(0.11) $$\sigma^*(x, \xi, y, \eta, t) = (1 - t)\,\sigma(x, \xi) + t\,\sigma(y, \eta)$$

then we have for $k = 1$ and $k = 2$ the formula

(0.12) $$c_k = \frac{(-i)^k}{2k}\left(\frac{1}{2\pi}\right)^n \int_{\mathbb{R}^n}\int_{\mathbb{R}^n}\int_0^1 (\nabla_x \nabla_\eta)^k f(\sigma^*)\Big|_{\substack{y=x\\ \eta=\xi}} \frac{dt}{t(1-t)}\, dx\, d\xi,$$

where now the integral does not have to be regularized. Here $\nabla_x \nabla_\eta$ denotes the operator

$$\sum_{i=1}^{n} \frac{\partial^2}{\partial x_i\, \partial \eta_i}.$$

This formula for $k = 1$ was not hard to discover; that is holds also for $k = 2$ we found utterly surprising. For $k > 2$ the formula is incorrect and the domain of the Symbol very likely has to be enlarged.

II. In [B] Basor found a formula for the third term in the asymptotic formula (0.2) when φ is piecewise smooth. If we write the expansion in the form

(0.13) $$\operatorname{tr} f(T_n(\varphi)) = a_1 n + a_2 \log n + a_3 + o(1)$$

then of course $a_1 = (2\pi)^{-1} \int f(\varphi(\theta))\, d\theta$ and a_2 is given by (0.3). The constant a_3 was itself written as a sum $a_3(1) + a_3(2)$. The formula for $a_3(1)$ in [B] can be stated in Symbolic terms by defining a Symbol φ^* with domain $\mathbb{T} \times \mathbb{T} \times [0, 1]$ by

$$\varphi^*(\theta, \psi, t) = (1 - t)\,\varphi(\theta-) + t\,\varphi(\psi+).$$

(In the smooth case, of course, this is the same as (0.7). The restriction of this φ^* to the diagonal of $\mathbb{T} \times \mathbb{T}$ is exactly the Gohberg-Krupnik Symbol (0.1).) In these terms the formula for $a_3(1)$ can be written

(0.14) $$\frac{1}{2\pi^2}\sum_{k=1}^{\infty} \int_0^1 \left[k \int_{\mathbb{T}}\int_{\mathbb{T}} e^{ik(\theta-\psi)} f(\varphi^*(\theta, \psi, t))\, d\theta\, d\psi - \frac{2}{k}\sum_{\theta} f(\varphi^*(\theta, \theta, t))\right] \frac{dt}{t(1-t)}$$

where the integral over t is regularized as usual. The sum in the integrand is taken over the points of discontinuity of φ, the terms corresponding to other points contributing zero.

The second constituent, $a_3(2)$, of a_3 is given in [B] by an infinite series of integrals of increasing order. Here we shall show that it is also representable as the regularized integral

(0.15) $$\frac{1}{2\pi^2}\int_0^1 \sum_{\theta} f(\varphi^*(\theta, \theta, t))\, u(t)\, dt$$

where

$$u(t) = \mathcal{Re}\, \frac{\Gamma'\left(\frac{1}{2} + \frac{1}{2\pi i}\log\frac{1-t}{t}\right)}{\Gamma\left(\frac{1}{2} + \frac{1}{2\pi i}\log\frac{1-t}{t}\right)} \Big/ t(1-t).$$

So once again the measure (0.5) appears, but this time with a mysterious factor involving the gamma function. (Stirling's formula shows that this factor behaves like log log $1/t(1-t)$ near $t = 0, 1$.)

Notice that if φ is smooth then the diagonal part of φ^* does not contribute at all and it is easily seen that (0.14) reduces to (0.6).

III. The finite Wiener-Hopf operator on $L_2(0, \alpha)$ with symbol $\sigma(\xi)$ is unitarily equivalent to the operator on $L_2(0, 1)$ with kernel given by (0.8) with $n = 1$ and $\sigma(x, \xi)$ replaced by $\sigma(\xi)$. In [W4] we considered the generalization of this to operators on $L_2(\Omega)$, where Ω is a bounded subset of $\mathbb{R}^n$ with smooth boundary, with σ in (0.8) allowed to depend on x also. Here, also, there was shown to be an expansion (0.9), but with coefficients given by much more complicated formulas than in the situation described in I. In [W5] we considered the even more general case of operators on $L_2(\mathbb{R}^n)$ with kernel given by (0.8) but where σ jumps across a hypersurface $\mathfrak{M}$ in x-space. (For the operators on $L_2(\Omega)$ described just above we would have a symbol which is smooth for $x \in \Omega$ and vanishes for $x \notin \Omega$; thus $\mathfrak{M}$ in this case is $\partial\Omega$.) Again it was shown—but this time only for polynomial f—that there is an expansion (0.9). Of course c_0 is given by (0.10) but now c_1 has two constituents. The first is just (0.12) with $k = 1$, the x-integration taken over the complement of $\mathfrak{M}$. For the second we shall here define a Symbol σ^* with domain $T^*(\mathfrak{M}) \times \mathbb{R} \times \mathbb{R} \times [0, 1]$, where $T^*(\mathfrak{M})$ is the cotangent bundle of $\mathfrak{M}$, and show that this constituent is obtained by applying a certain distribution (essentially another regularized integral) to $f(\sigma^*)$. In the finite Wiener-Hopf case this reduces to (the continuous analogue of) formula (0.6).

IV. Also in [W5] we considered the case where in (0.8) the symbol $\sigma(x, \xi)$ jumps across a hyperplane $\mathfrak{M}$ in (x, ξ)-space, and is otherwise smooth and rapidly decreasing. It was shown, again only for polynomial f, and for generic $\mathfrak{M}$, that there is then an expansion

$$\text{tr}\, f(T) \sim \sum_{k=0}^{\infty} c_k\, \alpha^{n-k/2}. \tag{0.16}$$

Of course c_0 is given by (0.10). We shall show that if we define σ^* on $\mathfrak{M} \times [0, 1]$ by

$$\sigma^*(x, \xi, t) = (1-t)\,\sigma^+(x, \xi) + t\,\sigma^-(x, \xi) \tag{0.17}$$

where $\sigma^{\pm}(x, \xi)$ are the limits of σ from the two sides of $\mathfrak{M}$, then c_1 is equal to the regularized integral

$$c_1 = \frac{\beta}{(2\pi)^n} \int_{\mathfrak{M}} \int_0^1 f(\sigma^*(x, \xi, t))\, w(t)\, dt\, d(x, \xi) \tag{0.18}$$

where β is a constant depending on the orientation of $\mathfrak{M}$, $d(x, \xi)$ denotes Lebesgue measure on $\mathfrak{M}$, and

$$w(t) = \frac{dE^{-1}(t)}{dt}$$

where E^{-1} is the inverse of the incomplete error functions given by

$$E(s) = \int_{-\infty}^{s} e^{-\pi r^2}\, dr\,. \tag{0.19}$$

(This is slightly different from the usual incomplete error function *Erf*.) Thus the measure $w(t)\, dt$ is just the push-forward to $[0, 1]$ of Lebesgue measure on $\mathbb{R}$ via the map $E : \mathbb{R} \to [0, 1]$. The weight function $w(t)$ is more singular than $1/t(1-t)$ by a factor of order $[\log 1/t(1-t)]^{1/2}$. Why it arises is another mystery.

Although the results just described fit the same general pattern their proofs are completely different.

For I we take the existing formulas for c_1 and c_2 and check that (0.12) holds. The check for $k = 1$ is trivial, for $k = 2$ not completely so.

In II, the derivation of formula (0.15) for the second constituent of a_3, we modify Basor's argument by using the functional equation for the Barnes G-function to evaluate a certain integral rather than its infinite product representation.

Section III is by far the longest of the paper and we derive our formula for the second constituent of c_1 (when σ in (0.8) jumps across a hypersurface in x-space) by a very circuitous route. Using the result in [W5] we show that it suffices to prove the correctness of the formula in the case $n = 1$. This case is done from the very beginning and requires two steps. First we show that c_1 has two constituents, that the first is what it ought to be, and that the second is equal to the trace of a certain operator associated with f and a singular integral operator on $L_2(\mathbb{R})$. Second, we find a formula for this trace with the help of the inversion of singular integral operators via Wiener-Hopf factorization. The reader may wonder why we cannot use the result of [W5] directly. The problem is that what is derived there is a formula for c_1 in the case where $f(\lambda) = \lambda^m$ with positive integer m and this formula involves an m-fold integral. Thus the formula increases in complexity as m increases and what we need is one in which m appears in a more benign way. There ought to be a more direct route from one to the other, but we do not see it.

For IV, the case where σ in (0.8) jumps across a hyperplane in (x, ξ)-space, we do use the formulas found in [W5] for the traces of the powers. They also increase in complexity as the power increases and involve constants given by integrals

$$\int (q\text{'th smallest component of } \omega)\, d\omega$$

where, for the m'th power, the integrals with $q = 1,\ldots, m$ are taken over the unit sphere in the subspace of $\mathbb{R}^m$ defined by $\Sigma\omega_i = 0$. We shall show here that these integrals can be represented nicely in terms of the error function $E(s)$ and this representation will lead to the expression (0.18) for c_1.

The four parts of the paper, which follow, can be read independently of each other.

I. SMOOTH SYMBOLS ON $\mathbb{R}^n$

Here is how (0.9) is obtained. The trace of a pseudodifferential operator (under, as usual, suitable conditions) equals $(2\pi)^{-n}$ times the integral of its symbol over $\mathbb{R}^n \times \mathbb{R}^n$. Our operator T with kernel (0.8) has symbol $\sigma(x, \xi/\alpha)$ and $f(T)$ is a pseudodifferential operator whose symbol has an asymptotic expansion

$$\sum_{k=0}^{\infty} \alpha^{-k}\, \gamma_k(x, \xi/\alpha) \tag{1.1}$$

with each γ_k a Schwartz function. Integrating (1.1) term-by-term, introducing the variable change $\xi \to \alpha\xi$, gives (0.9). The first few γ_k are given by

$$\gamma_0 = f(\sigma)$$
$$\gamma_1 = -\frac{i}{2} \nabla_x\sigma\ \nabla_\xi\sigma f''(\sigma)$$

and $\gamma_2 = \gamma_{2,2} + \gamma_{2,3} + \gamma_{2,4}$ where

$$\begin{cases} \gamma_{2,2} = -\frac{1}{4} \nabla_x^2\sigma\ \nabla_\xi^2\sigma f''(\sigma) \\ \gamma_{2,3} = -\frac{1}{6}\left[\nabla_x^2\sigma\ (\nabla_\xi\sigma)^2 + \nabla_x\sigma\ (\nabla_x\nabla_\xi\sigma)\ \nabla_\xi\sigma + (\nabla_x\sigma)^2\ \nabla_\xi^2\sigma\right] f'''(\sigma) \\ \gamma_{2,4} = -\frac{1}{8} (\nabla_x\sigma)^2\ (\nabla_\xi\sigma)^2 f^{(iv)}(\sigma). \end{cases} \tag{1.2}$$

Here we have used tensor notation so that, for example,

$$\nabla_x^2\sigma\ (\nabla_\xi\sigma)^2 = \sum_{i,j} \frac{\partial^2\sigma}{\partial x_i\, \partial x_j} \frac{\partial\sigma}{\partial\xi_i} \frac{\partial\sigma}{\partial\xi_j}$$

$$\nabla_x\sigma\ (\nabla_x\nabla_\xi\sigma)\ \nabla_\xi\sigma = \sum_i \frac{\partial^2\sigma}{\partial x_i\, \partial\xi_i} \cdot \sum_j \frac{\partial\sigma}{\partial x_j} \frac{\partial\sigma}{\partial\xi_j}.$$

(These formulas, in the context of compact manifolds, can be found in [W1, Sec. 5]. It is simpler in $\mathbb{R}^n$.)

It follows immediately that c_0 is given by (0.10) and c_1 by

$$-\frac{i}{2}\left(\frac{1}{2\pi}\right)^n \int\int \nabla_x\sigma\, \nabla_\xi\sigma\, f''(\sigma)\, dx\, d\xi.$$

This clearly equals the right side of (0.12) when $k = 1$. As for $k = 2$, the right side of (0.12) in this case is easily computed to be $-(2\pi)^{-n}/4$ times the integral over $\mathbb{R}^n \times \mathbb{R}^n$ of

(1.3) $$\nabla_x^2\sigma\, \nabla_\xi^2\sigma\, f''(\sigma) + \frac{1}{2}\left[\nabla_x^2\sigma\, (\nabla_\xi\sigma)^2 + (\nabla_x\sigma)^2\nabla_\xi\sigma\right] f'''(\sigma) + \frac{1}{6}\nabla_x\sigma\, \nabla_\xi\sigma\, f^{(iv)}(\sigma).$$

The terms here are in general like those in (1.2) except for the term

(1.4) $$\nabla_x\sigma\, (\nabla_x\nabla_\xi\sigma)\, \nabla_\xi\sigma\, f'''(\sigma)$$

in the expression for $\gamma_{2,3}$ which does not appear in (1.3). However since we are eventually going to integrate over $\mathbb{R}^n \times \mathbb{R}^n$, integration by parts allows us to make a replacement for this term. Removing the ∇_x from the second factor in (1.4) and distributing it among the others (and remembering to change sign) shows that (1.4) can be replaced by

$$-\nabla_x^2\sigma\, (\nabla_\xi\sigma)^2 f'''(\sigma) - \nabla_x\sigma\, \nabla_\xi\sigma\, \nabla_x\nabla_\xi\sigma\, f'''(\sigma) - (\nabla_x\sigma)^2\, (\nabla_\xi\sigma)^2 f^{(iv)}(\sigma).$$

Noting that the middle term here equals minus (1.4) itself, we see that (1.4) can also be replaced by

$$-\frac{1}{2}\left[\nabla_x^2\sigma\, (\nabla_\xi\sigma)^2 f'''(\sigma) + (\nabla_x\sigma)^2\, (\nabla_\xi\sigma)^2 f^{(iv)}(\sigma)\right].$$

By symmetry in x and ξ, (1.4) can also be replaced by the symmetrization of this over x and ξ, which is

$$-\frac{1}{4}\left[\nabla_x^2\sigma\, (\nabla_\xi\sigma)^2 + (\nabla_x\sigma)^2\, \nabla_\xi^2\sigma\right] f'''(\sigma) - \frac{1}{2}(\nabla_x\sigma)^2\, (\nabla_\xi\sigma)^2 f^{(iv)}(\sigma).$$

Making this replacement for the term (1.4) appearing in the expression for $\gamma_{2,3}$ we find that the integral of γ_2 is equal to $-1/4$ times the integral of (1.3) and this establishes (0.12) when $k = 2$. A miracle?

II. PIECEWISE SMOOTH SYMBOLS ON $\mathbb{T}$

A word about the derivation of (0.13). With integration over an appropriate contour, we have

$$\operatorname{tr} f(T_n(\varphi)) = \frac{1}{2\pi i}\int f(\lambda)\, \operatorname{tr}\, (\lambda - T_n(\varphi))^{-1}\, d\lambda$$

$$= \frac{1}{2\pi i}\int f(\lambda)\, \frac{d}{d\lambda} \log\det\, (\lambda - T_n(\varphi))\, d\lambda \;= -\frac{1}{2\pi i}\int f'(\lambda)\, \log\det\, T_n(\varphi - \lambda)\, d\lambda.$$

One inserts the known asymptotics of Toeplitz determinants with piecewise smooth symbol into the integral and evaluates the result. It is here that the Barnes G-function arises. (We refer the reader to [BS, Chap. 10] for a discussion of these things.)

By this method the formula for a_3 was derived in [B]. With $a_3 = a_3(1) + a(2)$ the expression for $a_3(1)$ was expressible as (0.14). We shall not state the final formula for $a_3(2)$ derived in [B] but only that it is equal to

$$-\frac{1}{2\pi i}\int f'(\lambda)\sum_{\theta}\log g\left(\frac{1}{2\pi i}\log\frac{\lambda-\varphi(\theta+)}{\lambda-\varphi(\theta-)}\right)d\lambda$$

where the sum is taken over the points of discontinuity of φ and where g is related to the Barnes G-function by

$$g(\lambda) = G(1+\lambda)\,G(1-\lambda).$$

The logarithm appearing in the argument of g is determined by saying that it vanishes when $\lambda = \infty$.

Consider a term corresponding to a single discontinuity θ, and for convenience set $a = \varphi(\theta-)$, $b = \varphi(\theta+)$. Thus we consider

$$-\frac{1}{2\pi i}\int f'(\lambda)\,\log g\left(\frac{1}{2\pi i}\log\frac{\lambda-b}{\lambda-a}\right)d\lambda\,. \tag{2.1}$$

Now G is an entire function with zeros at the nonpositive integers, and so g is entire with zeros at the nonzero integers. The argument of g in the integral can never be an integer since $a \neq b$. Moreover at $\lambda = \infty$ this argument equals 0, and $g(0) \neq 0$. It follows that the log g term in the integral is single-valued and analytic in the λ-plane cut along the line segment joining $\lambda = a$ to $\lambda = b$. The path of integration can be replaced by this cut, transversed twice in opposite directions, with the limiting values of log g from the two sides of the cut in the integrand.

If we face in the direction from a to b then the limiting value of the argument of g as λ approaches the cut from the left is

$$\frac{1}{2}+\frac{1}{2\pi i}\log\left|\frac{\lambda-b}{\lambda-a}\right| = \frac{1}{2}+\frac{1}{2\pi i}\log\frac{b-\lambda}{\lambda-a}\,;$$

for the limiting value from the right of the cut we change the summand 1/2 to $-1/2$. It follows that (2.1) equals

$$\frac{1}{2\pi i}\int_a^b f'(\lambda)\,\log\frac{g\left(\frac{1}{2}+\frac{1}{2\pi i}\log\frac{b-\lambda}{\lambda-a}\right)}{g\left(-\frac{1}{2}+\frac{1}{2\pi i}\log\frac{b-\lambda}{\lambda-a}\right)}\,d\lambda\,.$$

Now the functional equation for the G-function is $G(z+1) = \Gamma(z)G(z)$ and it follows that

$$\frac{g(z)}{g(z-1)} = \frac{\Gamma(z)}{\Gamma(1-z)}\,.$$

Hence the logarithm of the quotient in the integral equals

$$\log \frac{\Gamma\left(\frac{1}{2}+\frac{1}{2\pi i}\log\frac{b-\lambda}{\lambda-a}\right)}{\Gamma\left(\frac{1}{2}-\frac{1}{2\pi i}\log\frac{b-\lambda}{\lambda-a}\right)}\,.$$

This equals

$$2i\,\mathcal{Im}\,\log\Gamma\left(\frac{1}{2}+\frac{1}{2\pi i}\log\frac{b-\lambda}{\lambda-a}\right)$$

since $\Gamma(z)$ takes conjugate values at conjugate points and $(b-\lambda)/(\lambda-a)$ is positive real for λ on the cut. Hence if we parametrize the cut by

$$\lambda=(1-t)\,a+t\,b \qquad\qquad 0\le t\le 1$$

our integral becomes

$$\frac{1}{\pi}\int_0^1 \frac{d}{dt}f((1-t)\,a+t\,b)\;\mathcal{Im}\,\log\Gamma\left(\frac{1}{2}+\frac{1}{2\pi i}\log\frac{1-t}{t}\right)dt\,.$$

Before integrating by parts we observe that if the factor involving f were replaced by a constant the result would be zero. This is shown by making the variable change $t\to 1-t$ and using once again the fact that $\Gamma(z)$ takes conjugate values at conjugate points. So we can replace $f((1-t)\,a+t\,b)$ by

(2.2) $$f((1-t)\,a+t\,b)-[(1-t)\,f(a)+t\,f(b)]$$

in the integral and then integrate by parts. (This replacement is important since (2.2) vanishes at $t=0,1$ and the Γ factor in the integral is mildly singular there.) What results is

$$\frac{1}{2\pi^2}\int_0^1 \left\{f((1-t)\,a+t\,b)-[(1-t)\,f(a)+t\,f(b)]\right\}\mathcal{Re}\frac{\Gamma'\left(\frac{1}{2}+\frac{1}{2\pi i}\log\frac{1-t}{t}\right)}{\Gamma\left(\frac{1}{2}+\frac{1}{2\pi i}\log\frac{1-t}{t}\right)}\,\frac{dt}{t(1-t)}\,.$$

Recalling that we are to replace a by $\varphi(\theta-)$ and b by $\varphi(\theta+)$ and then sum over the discontinuities θ we see that the representation (0.15) for $a_3(2)$ has been established.

III. PIECEWISE SMOOTH SYMBOLS ON $\mathbb{R}^n$

In this section, for later convenience of notation, we replace ξ in (0.8) by ζ and reserve ξ for a variable running over $\mathbb{R}$. We are given a hypersurface $\mathcal{M}$ in x-space and assume that $\sigma(x,\zeta)$ is C^∞ in the complement of $\mathcal{M}\times\mathbb{R}^n$, with all derivatives bounded and rapidly decreasing as $|x|+|\zeta|\to\infty$. It was shown in [W5] that under a uniform smoothness condition on $\mathcal{M}$ and for polynomial f vanishing at 0 there is an expansion (0.9) with c_0 given by (0.10). The formula for c_1 had two parts, and so we write here $c_1=c_1(1)+c_1(2)$. The first summand $c_1(1)$ is just (0.12), the integral being taken over the complement of $\mathcal{M}\times\mathbb{R}^n$. The formula for $c_1(2)$, in the case $f(\lambda) = \lambda^m$, involves a multiple integral whose order increases with m. We here derive a completely different expression for $c_1(2)$, and we begin by introducing some notation.

For $x \in \mathfrak{M}$ we denote by ν_x a unit normal vector to $\mathfrak{M}$ at x and define, first

$$\sigma^{\pm}(x, \zeta) = \lim_{s \to 0\pm} \sigma(x + s\nu_x, \zeta), \tag{3.1}$$

and then for $\zeta \in \nu_x^{\perp}$,

$$\sigma^{\pm}_{(x,\zeta)}(\xi) = \sigma^{\pm}(x, \zeta + \xi\nu_x), \qquad \xi \in \mathbb{R}.$$

The pair (x, ζ) is a point of the cotangent bundle $T^*(\mathfrak{M})$ and we integrate over it by the formula $\int_{\mathfrak{M}} \int_{\nu_x^{\perp}} \cdots d\zeta\, dx$, with $d\zeta$ denoting Lebesgue measure on $\nu_x^{\perp}$ and dx surface measure on $\mathfrak{M}$.

We associate with an ordered pair of functions (σ_1, σ_2) on $\mathbb{R}$ a 2×2 matrix of functions $(\sigma_1, \sigma_2)^*_{i,j}$ on $\mathbb{R} \times \mathbb{R} \times [0, 1]$ as follows:

$$(3.2)\qquad \begin{aligned} &(\sigma_1, \sigma_2)^*_{i,j}(\xi, \eta, t) = (1 - t)\, \sigma_i(\xi) + t\, \sigma_j(\eta) \qquad \text{if } i, j \neq 2, 1, \\ &(\sigma_1, \sigma_2)^*_{2,1} = (\sigma_1, \sigma_2)^*_{1,2}\,. \end{aligned}$$

Here of course i and j take the values 1 and 2. The matrix is symmetric but depends on the order of the pair of functions.

Our Symbol σ^* is defined to be the matrix-valued function on $T^*(\mathfrak{M}) \times \mathbb{R} \times \mathbb{R} \times [0, 1]$ whose value at the point $(x, \zeta), \xi, \eta, t$ is obtained from (3.2) by setting

$$\sigma_1 = \sigma^{+}_{(x,\zeta)},\ \sigma_2 = \sigma^{-}_{(x,\zeta)}\,.$$

(This Symbol, appropriate for the representation of $c_1(2)$, is of course different from the σ^* defined by (0.11) which is appropriate for the representation of $c_1(1)$.) With these definitions the formula we shall establish for the second constituent of c_1 is

$$(3.3)\quad c_1(2) = \frac{1}{2}\left(\frac{1}{2\pi}\right)^{n+1} \int_{T^*(\mathfrak{M})} \int_{\mathbb{R}} \int_{\mathbb{R}} \int_0^1 \sum_{i,j=1}^{2} (-1)^{i+j} f(\sigma^*_{i,j})\, \frac{dt}{t(1-t)}\, \frac{d\eta\, d\xi}{(\xi - \eta + 0i)^2}\, d\zeta\, dx$$

where, as usual, the integral is regularized by subtracting from $f(\sigma^*_{i,j})$ its linear part. The integral over η is interpreted as the limit as $\varepsilon \to 0+$ of the corresponding integral with $0i$ replaced by εi.

Here is why the result of [W5] allows us to reduce the proof of this to the case $n = 1$. The expression for $c_1(2)$ obtained there can be cast into the form

$$\left(\frac{1}{2\pi}\right)^{n+1} \int_{T^*(\mathfrak{M})} F_f\,(\sigma^{+}_{(x,\zeta)}, \sigma^{-}_{(x,\zeta)})\, d\zeta\, dx$$

where F_f is a function, depending on the polynomial f but not the dimension n, on $\mathcal{S}(\mathbb{R}) \times \mathcal{S}(\mathbb{R})$. (Here $\mathcal{S}(\mathbb{R})$ denotes the Schwartz class on $\mathbb{R}$.) If we can prove (3.3) when $n = 1$ and $\mathfrak{M} = \{0\}$ then we will have enough information to identify $F_f\,(\sigma_1, \sigma_2)$ as equal to

$$\frac{1}{8\pi^2}\int_0^1 \int_{\mathbb{R}} \int_{\mathbb{R}} \sum_{i,j=1}^{2} (-1)^{i+j} f((\sigma_1, \sigma_2)^*_{i,j}) \frac{dt}{t(1-t)} \frac{d\eta\, d\xi}{(\xi - \eta + 0i)^2} \tag{3.4}$$

and so (3.3) will follow for general n.

From now on we take $n = 1$, $\mathfrak{M} = \{0\}$, $\nu_x = 1$ and, since there is only the single point $(0, 0)$ in $T^*(\mathfrak{M})$, we drop the subscript (x, ζ). Thus

$$\sigma^{\pm}(\xi) = \lim_{x \to 0\pm} \sigma(x, \xi).$$

We shall prove two lemmas which, when combined, give the desired result. In them we use the notation $P_{\pm}$ for multiplication by the characteristic function of $\mathbb{R}^{\pm}$ and op $\sigma(x, \xi)$ for the pseudodifferential operator with symbol $\sigma(x, \xi)$, the operator whose kernel is given by (0.8) with $\alpha = 1$. Also, we denote by $\sigma^{\pm}(x, \xi)$ any functions on $\mathcal{S}(\mathbb{R} \times \mathbb{R})$ which agree with $\sigma(x, \xi)$ for $x \in \mathbb{R}^{\pm}$. (Of course these have no connection with the functions in (3.1).) Thus $\sigma^{\pm}(\xi) = \sigma^{\pm}(0, \xi)$. In these terms we can write

$$T = P_+ \operatorname{op} \sigma^+(x, \xi/\alpha) + P_- \operatorname{op} \sigma^-(x, \xi/\alpha)$$

and we shall prove the following.

LEMMA A. *If f is a polynomial vanishing at 0 we have as $\alpha \to \infty$*

$$\operatorname{tr} f(T) = c_0\, \alpha + c_1(1) + c_1(2) + o(1)$$

where

$$c_0 = \frac{1}{2\pi} \int \int f(\sigma(x, \xi))\, dx\, d\xi$$

$$c_1(1) = -\frac{i}{2} \frac{1}{2\pi} \int \int \nabla_x \sigma\, \nabla_\xi \sigma\, f''(\sigma)\, dx\, d\xi$$

$$c_1(2) = \operatorname{tr}\left[f(P_+ \operatorname{op} \sigma^+ + P_- \operatorname{op} \sigma^-) - P_+ \operatorname{op} f(\sigma^+) - P_- \operatorname{op} f(\sigma^-)\right]. \tag{3.5}$$

(In the last, $\sigma^{\pm}$ denote the functions $\sigma^{\pm}(\xi)$, which are independent of x.)

LEMMA B. *For any f which is analytic on a convex set containing the ranges of the Schwartz functions $\sigma_1(\xi)$ and $\sigma_2(\xi)$ the trace*

$$\operatorname{tr}\left[f(P_+ \operatorname{op} \sigma_1 + P_- \operatorname{op} \sigma_2) - P_+ \operatorname{op} f(\sigma_1) - P_- \operatorname{op} f(\sigma_2)\right]$$

is equal to (3.4).

Clearly Lemmas A and B yield (3.3) in the case $n = 1$, $\mathfrak{M} = \{0\}$. For the proof of Lemma A, which will require some sublemmas, we use the fact that the operators op $\sigma^{\pm}(x, \xi/\alpha)$

and op $\sigma^{\pm}(x/\alpha, \xi)$ are unitarily equivalent (via the unitary map $\varphi(x) \to \alpha^{1/2}\,\varphi(\alpha x)$ of $L_2(\mathbb{R})$) and so for each power m we have

(3.6) $$\operatorname{tr} T^m = \operatorname{tr}([P_+ \operatorname{op} \sigma^+(x/\alpha, \xi) + P_- \operatorname{op} \sigma^-(x/\alpha, \xi)]^m).$$

In what follows we denote by $\tau, \tau_1, \ldots$ any functions in $\mathcal{S}(\mathbb{R} \times \mathbb{R})$.

SUBLEMMA 1. (i) *As* $\alpha \to \infty$ *the operator* op $\tau(x/\alpha, \xi)$ *converges strongly to the operator* op $\tau(0, \xi)$. (ii) *The operator* $P_+ \operatorname{op} \tau(x/\alpha, \xi)\, P_- - P_+ \operatorname{op} \tau(0, \xi)\, P_-$ *converges to* 0 *in the Hilbert-Schmidt norm as* $\alpha \to \infty$, *and the same is true if* P_+ *and* P_- *are interchanged.*

PROOF. The kernel of op $\tau(x/\alpha, \xi)$ is $k(x/\alpha, x - y)$ where k is the inverse Fourier transform of τ in its second variable. This function converges pointwise to $k(0, x - y)$ as $\alpha \to \infty$ and is bounded by a constant times $\langle x - y\rangle^{-N}$, uniformly in α, for each N. (We use the notation $\langle x \rangle = (1 + x^2)^{1/2}$.) Assertion (i) follows quickly from this. As for (ii), the kernel of the operator in question is bounded by a constant times $|x/\alpha|\, \langle x - y\rangle^{-N}$, whence the square of its Hilbert-Schmidt norm is at most a constant times

$$\int_0^\infty \int_0^\infty (x/\alpha)^2 \langle x + y\rangle^{-2N}\, dy\, dx$$

which is $O(\alpha^{-2})$ if $N > 2$.

In the next two sublemmas we use the notation $T_1 \equiv T_2$ for operators T_i depending on the parameter α if $\operatorname{tr}(T_1 - T_2) \to 0$ as $\alpha \to \infty$.

SUBLEMMA 2. *If each operator* R_i *is either* P_+ or P_-, *and some* R_i *is* P_-, *then*

$$P_+ \operatorname{op} \tau_1(x/\alpha, \xi)\, R_1 \operatorname{op} \tau_2(x/\alpha, \xi)\, R_2 \cdots R_{m-1} \operatorname{op} \tau_m(x/\alpha, \xi)\, P_+$$

(3.7)

$$\equiv P_+ \operatorname{op} \tau_1(0, \xi)\, R_1 \operatorname{op} \tau_2(0, \xi)\, R_2 \cdots R_{m-1} \operatorname{op} \tau_m(0, \xi)\, P_+.$$

The same holds when P_+ *and* P_- *are interchanged.*

PROOF. In the product on the left appear factors

$$P_+ \operatorname{op} \tau_i(x/\alpha, \xi)\, P_- \text{ and } P_- \operatorname{op} \tau_j(x/\alpha, \xi)\, P_+.$$

By Sublemma 1 (ii) these converge in Hilbert-Schmidt norm to

$$P_+ \operatorname{op} \tau_i(0, \xi)\, P_- \text{ and } P_- \operatorname{op} \tau_j(0\, \xi)\, P_+.$$

By Sublemma 1 (i) the other factors $R_{k-1} \operatorname{op} \tau_k(x/\alpha, \xi) R_k$ converge strongly to $R_{k-1} \operatorname{op} \tau_k(0, \xi) R_x$. It follows by standard operator theory that the operator on the left in (3.7) converges in trace norm to the operator on the right.

In the following recall that $\sigma^{\pm}(\xi) = \sigma^{\pm}(0, \xi)$ and that we write op $\sigma^{\pm}$ for op $\sigma^{\pm}(\xi)$.

SUBLEMMA 3. *For each positive integer m we have*

$$[P_+ \text{ op } \sigma^+(x/\alpha, \xi) + P_- \text{ op } \sigma^-(x/\alpha, \xi)]^m - [P_+ \text{ op } \sigma^+(x/\alpha, \xi)]^m - [P_- \text{ op } \sigma^-(x/\alpha, \xi)]^m$$
$$\equiv (P_+ \text{ op } \sigma^+ + P_- \text{ op } \sigma^-)^m - (P_+ \text{ op } \sigma^+)^m - (P_- \text{ op } \sigma^-)^m.$$

PROOF. Expand the m'th powers of the sums and apply Sublemma 2, using the general fact

$$\text{tr } P_\pm A = \text{tr } A\, P_\pm = \text{tr } P_\pm A\, P_\pm. \tag{3.8}$$

We should remark that, strictly speaking, (3.8) requires that the operator A be trace class and so we have used, for example, the fact that P_+ op $\sigma^+ P_-$ op σ^- is trace class. This is certainly true—in fact, P_+ op $\sigma^+ P_-$ is itself trace class. The issue can be avoided by thinking of the trace of an operator as the integral of its kernel over the diagonal when it is continuous almost everywhere on the diagonal.

SUBLEMMA 4. $(P_\pm \text{ op } \sigma^\pm(x/\alpha, \xi))^m - P_\pm (\text{op } \sigma^\pm(x/\alpha, \xi))^m \equiv (P_\pm \text{ op } \sigma^\pm)^m - P_\pm (\text{op } \sigma^\pm)^m$.

PROOF. Take the case of + signs. The left side, after inserting a factor P_+ to its right (which does not change its trace) can be written as a sum of expressions of the form of the left side of (3.7), and the right side as an exactly corresponding sum of expressions of the form of the right side of (3.7). Thus the conclusion follows from Sublemma 2.

SUBLEMMA 5. There is a complete asymptotic expansion

$$\text{tr}\left[P_+(\text{op } \sigma^+(x/\alpha, \xi))^m + P_-(\text{op } \sigma^-(x/\alpha, \xi))^m\right] \sim \sum_{k=0}^{\infty} a_k\, \alpha^{1-k}$$

with

$$a_0 = \frac{1}{2\pi} \int\int \sigma(x, \xi)^m\, dx\, d\xi, \quad a_1 = -\frac{im(m-1)}{2} \frac{1}{2\pi} \int\int \frac{\partial\sigma}{\partial x} \frac{\partial\sigma}{\partial\xi} \sigma^{m-2}\, dx\, d\xi$$

where the second integration is taken over the region where $x \neq 0$.

PROOF. We invoke, as we did in I, a fact about pseudodifferential operators, in this case that there is a representation

$$\text{op } \tau_1(x/\alpha, \xi) \text{ op } \tau_2(x/\alpha, \xi) = \text{op } \tau(x/\alpha, \xi)$$

where τ has the asymptotic expansion

$$\tau(x, \xi) \sim \sum_{k=0}^{\infty} \frac{(i\alpha)^{-k}}{k!} \partial_\xi^k \tau_1(x/\alpha, \xi)\, \partial_x^k \tau_2(x/\alpha, \xi).$$

This is the standard expansion for the symbol of a product [H, Th. 18.18], modified by the presence of the parameter α. What it means in this case is that the difference between $\tau(x, \xi)$ and the partial sum $\Sigma_{k<K}$ is of the form $\alpha^{-K}\,\tau_K(x/\alpha, \xi)$ with τ_K a Schwartz function. By induction there is an expansion for the symbol of $(\mathrm{op}\,\tau(x/\alpha, \xi))^m$ beginning

$$\tau(x/\alpha, \xi)^m - \alpha^{-1}\frac{i}{2}m(m-1)\,\partial_\xi\tau(x/\alpha, \xi)\,\partial_x\tau(x/\alpha, \xi)\,\tau(x/\alpha, \xi)^{m-2} + \cdots .$$

It follows that there are asymptotic expansions for the (discontinuous) symbols of $P_\pm(\mathrm{op}\,\sigma^\pm(x/\alpha, \xi))^m$ beginning

$$\chi_{\mathbb{R}^\pm}(x)\left[\sigma^\pm(x, \xi)^m - \alpha^{-1}\frac{i}{2}m(m-1)\,\partial_\xi\sigma^\pm(x/\alpha, \xi)\,\partial_x\sigma^\pm(x/\alpha, \xi)\,\sigma^\pm(x/\alpha, \xi)^{m-2} + \cdots\right].$$

From the symbol we obtain the kernel from (0.8) and integrate the kernel over the diagonal to get the trace.

If we combine Sublemmas 3-5 and use (3.6) and the fact that $(\mathrm{op}\,\sigma^\pm)^m = \mathrm{op}\,(\sigma^\pm)^m$ since $\sigma^\pm$ is independent of x, we see that Lemma A is proved.

If $\sigma_2 = 0$ in Lemma B then we are essentially dealing with Wiener-Hopf operators and the result was proved in [W3]. Our method here is similar to the one there but the ending, the derivation of the final formula, is a little different.

We begin with the case $f(\lambda) = \lambda^{-1}$. Of course σ_1 and σ_2 cannot then be Schwartz functions. We assume they are of the form constant (the same for both) plus Schwartz function. It follows from our assumption that the convex hulls of the ranges of the σ_i do not contain 0 that the σ_i have zero index, and hence have continuous logarithms taking the same values at $\xi = \pm\infty$. And it follows from this that σ_1/σ_2 has a continuous logarithm, which is in fact a Schwartz function. We use this logarithm to construct the Wiener-Hopf factorization of σ_1/σ_2.

We define the operators $\tau \to \tau_\pm$ by

(3.9) $$\tau_\pm(\xi) = \pm\frac{i}{2\pi}\int\frac{\tau(\eta)}{\eta - \xi \pm 0i}\,d\eta = \pm\frac{i}{2\pi}\left[PV\int\frac{\tau(\eta)}{\eta - \xi}\,d\eta \mp \pi i\,\sigma(\xi)\right].$$

Then

(3.10) $$\tau = \tau_+ + \tau_-$$

and τ_+ (resp. τ_-) extends to a bounded analytic function in the lower (resp. upper) half-plane. We define the Wiener-Hopf factors $\sigma^\pm$ (which have no relation to the functions $\sigma^\pm$ earlier in this section) by

(3.11) $$\sigma^+ = \exp\left(\log\frac{\sigma_1}{\sigma_2}\right)_+, \quad \sigma^- = \exp\left(\log\frac{\sigma_2}{\sigma_2}\right)_- .$$

Thus $\sigma_1/\sigma_2 = \sigma^+/\sigma^-$ and σ^+ (resp. σ^-) extends to be analytic, bounded, and bounded away from zero in the lower (resp. upper) half-plane. The inverse of our operator is given by

$$(P_+\,\mathrm{op}\,\sigma_1 + P_-\,\mathrm{op}\,\sigma_2)^{-1} = \mathrm{op}(\sigma_1\,\sigma^-)^{-1}\,P_+\,\mathrm{op}\,\sigma^- + \mathrm{op}(\sigma_2\,\sigma^+)^{-1}\,P_-\,\mathrm{op}\,\sigma^+.$$

This is an easy check, using the fact that op $\sigma^\pm$ leaves invariant the range of $P_\pm$. (Or see [GF, p. 129].)

Next we observe that

$$P_+ \operatorname{op} \sigma_1^{-1} + P_- \operatorname{op} \sigma_2^{-1} = P_+ \operatorname{op}(\sigma_1 \sigma^-)^{-1} \operatorname{op} \sigma^- + P_- \operatorname{op}(\sigma_2 \sigma^+)^{-1} \operatorname{op} \sigma^+$$

and so, subtracting,

$$\begin{aligned}(3.12)\qquad &(P_+ \operatorname{op} \sigma_1 + P_- \operatorname{op} \sigma_2)^{-1} - (P_+ \operatorname{op} \sigma_1^{-1} + P_- \operatorname{op} \sigma_2^{-1}) \\ &= [\operatorname{op}(\sigma_1 \sigma^-)^{-1}, P_+] \operatorname{op} \sigma^- + [\operatorname{op}(\sigma_2 \sigma^+)^{-1}, P_-] \operatorname{op} \sigma^+.\end{aligned}$$

Here we have used the usual bracket notation to denote commutators. It is the trace of the left side we are interested in.

We pause now for some notations and sublemmas. First we remark that the functions $\sigma^\pm$ appearing on the right side of (3.12) are no longer of the form constant plus Schwartz function, so we must enlarge the class of functions we consider. We choose $\mathbb{C} + S^-$. Here, of course, $\mathbb{C}$ represents the constant functions; S^- denotes the class of what are called symbols of negative order, those C^∞ functions τ in $\mathbb{R}$ which satisfy estimates

$$\tau^{(k)}(\xi) = O(\langle\xi\rangle^{-s-k}) \qquad k = 0, 1, \ldots$$

for some $s > 0$. The point is that the operators $\tau \to \tau_\pm$ send S^- to S^- and that, of course, $\mathcal{S}(\mathbb{R}) \subset S^-$. We extend these operators to $\mathbb{C} + S^-$ by defining them to vanish on $\mathbb{C}$.

The Fourier transform of a function in S^- is rapidly decreasing at ∞ and $O(|x|^{s-1})$ near $x=0$. The Fourier transform of a function in $\mathbb{C} + S^-$ has in addition a δ-summand. For $\tau_1, \tau_2 \in \mathbb{C} + S^-$ we define

$$(3.13)\qquad \langle \tau_1, \tau_2 \rangle = \int x\, \check{\tau}_1(-x)\, \check{\tau}_2(x)\, dx,$$

the integral taken over $\mathbb{R}\backslash\{0\}$, where

$$\check{\tau}(x) = \frac{1}{2\pi} \int e^{i\xi x}\, \tau(\xi)\, d\xi\, .$$

SUBLEMMA 1. *If* $\tau_1 \in \mathbb{C} + S^-$ *and* $\tau_2 \in S^-$ *then*

$$\operatorname{tr} [\operatorname{op} \tau_1, P_\pm] \operatorname{op} \tau_2 = \pm \langle \tau_1, \tau_2 \rangle.$$

PROOF. An easy check, using the fact that

$$[\operatorname{op} \tau_1, P_+] = P_- \operatorname{op} \tau_1 P_+ - P_+ \operatorname{op} \tau_1 P_-.$$

Here, incidentally, there is no guarantee that the operators are trace class. They do, however, have rapidly decreasing kernels, continuous on the diagonal except at $x = 0$, and the trace is to be interpreted as the integral of the kernel over the diagonal.

SUBLEMMA 2. *If* $\log \tau_i \in \mathbb{C} + S^-$ *then*

$$\langle \tau_1, \tau_2 \rangle = \frac{1}{2} \left\langle \tau_1 \tau_2, \log \frac{\tau_1}{\tau_2} \right\rangle .$$

PROOF. This is left as an exercise for the reader; the proof can be found on pp. 293-4 of [W3].

SUBLEMMA 3. *With the same assumptions as Sublemma 2 we have*

$$\left\langle \tau_1^{-1}, (\log \tau_2)_- \right\rangle - \left\langle \tau_2^{-1}, (\log \tau_1)_+ \right\rangle$$

$$= -\frac{1}{4\pi^2} \int_{\mathbb{R}} \int_{\mathbb{R}} \int_0^1 \tau^\dagger(\xi, \eta, t)^{-1} \frac{dt}{t(1-t)} \frac{d\eta\, d\xi}{(\xi - \eta + 0i)^2}$$

where

$$\tau^\dagger(\xi, \eta, t) = (1-t)\, \tau_1(\xi) + t\, \tau_2(\eta)$$

and the integral is regularized by subtracting from $\tau^\dagger(\xi, \eta, t)^{-1}$ *its linear part.*

PROOF. We begin with another representation of our bilinear form,

$$\langle \tau_1, \tau_2 \rangle = \frac{i}{2\pi} \int \tau_1(\xi)\, \tau_2'(\xi)\, d\xi,$$

which follows from (3.13) using Parseval's identity. It follows from this, and the definition (3.9), that

$$\left\langle \tau_1^{-1}, (\log \tau_2)_- \right\rangle = \frac{1}{4\pi^2} \int \int \tau_1(\xi)^{-1} \log \tau_2(\eta) \frac{d\eta\, d\xi}{(\xi - \eta + 0i)^2},$$

$$\left\langle \tau_2^{-1}, (\log \tau_1)_+ \right\rangle = -\frac{1}{4\pi^2} \int \int \tau_2(\xi)^{-1} \log \tau_1(\eta) \frac{d\eta\, d\xi}{(\xi - \eta - 0i)^2}.$$

In the second integral we interchange the variables ξ and η and then the order of integration (which can be justified). We obtain

$$\left\langle \tau_1^{-1}, (\log \tau_2)_- \right\rangle - \left\langle \tau_2^{-1}, (\log \tau_1)_+ \right\rangle$$

$$= \frac{1}{4\pi^2} \int \int \left[\frac{\log \tau_2(\eta)}{\tau_1(\xi)} + \frac{\log \tau_1(\xi)}{\tau_2(\eta)} \right] \frac{d\eta\, d\xi}{(\xi - \eta - 0i)^2}. \tag{3.14}$$

The trick now is to apply the readily verified identity

$$\int_0^1 \left\{ ((1-t)\, a + t\, b)^{-1} - [(1-t)\, a^{-1} + t\, b^{-1}] \right\} \frac{dt}{t(1-t)} = \left(\frac{1}{a} - \frac{1}{b} \right) (\log a - \log b)$$

with $a = \tau_1(\xi)$ and $b = \tau_2(\eta)$. Two of the four terms obtained by multiplying out the right side here give exactly the negative of expression in brackets in the integrand in (3.14). The analogous integral with the other two terms

$$\int\int \left[\frac{\log \tau_1(\xi)}{\tau_1(\xi)} + \frac{\log \tau_2(\eta)}{\tau_2(\eta)}\right] \frac{d\eta\, d\xi}{(\xi - \eta + 0i)^2},$$

vanishes since, for example,

$$\int \frac{d\eta}{(\xi - \eta + 0i)^2} = 0$$

for all ξ. Thus we see that the sublemma has been established.

We can now quickly finish the proof of Lemma B. First, it follows from (3.12) and Sublemmas 1 and 2 that

$$\operatorname{tr}\left[(P_+ \operatorname{op} \sigma_1 + P_- \operatorname{op} \sigma_2)^{-1} - (P_+ \operatorname{op} \sigma_1^{-1} + P_- \operatorname{op} \sigma_2^{-1})\right]$$
$$= \frac{1}{2}\left\langle \sigma_1^{-1}, \log \sigma_1(\sigma^-)^2 \right\rangle - \frac{1}{2}\left\langle \sigma_2^{-1}, \log \sigma_2(\sigma^+)^2 \right\rangle.$$

If we use the definition (3.11) of $\sigma^\pm$ and identity (3.10) we find after a little computation that this equals

$$\left\langle \sigma_1^{-1}, (\log \sigma_2)_- \right\rangle - \left\langle \sigma_2^{-1}, (\log \sigma_1)_+ \right\rangle$$
$$+ \frac{1}{2}\left\langle \sigma_1^{-1}, (\log \sigma_1)_+ - (\log \sigma_1)_- \right\rangle + \frac{1}{2}\left\langle \sigma_2^{-1}, (\log \sigma_2)_+ - (\log \sigma_2)_- \right\rangle.$$

We apply Sublemma 3 with $\tau_1 = \sigma_1$, $\tau_2 = \sigma_2$ to evaluate the first two terms here, the sublemma with $\tau_1 = \tau_2 = \sigma_1$ to evaluate the third term, and the sublemma with $\tau_1 = \tau_2 = \sigma_2$ to evaluate the fourth. What results is (3.4) with $f(\lambda) = \lambda^{-1}$.

Thus Lemma B is proved in this special case. To prove it in general we apply this result, with σ replaced by $\lambda - \sigma$, then multiply by $f(\lambda)$, and integrate with respect to λ over the usual contour.

IV. SYMBOLS DISCONTINUOUS ACROSS A HYPERPLANE IN $\mathbb{R}^n \times \mathbb{R}^n$

Now we assume that σ in (0.8) is smooth in the complement of a hyperplane $\mathfrak{M}$ in $\mathbb{R}^n \times \mathbb{R}^n$, and that all derivatives are bounded and rapidly deceasing as $|x| + |\xi| \to \infty$, and we derive the formula (0.18) for the coefficient c_1 in the expansion (0.16) from the formula in [W5]. To state it, we denote by ν a unit vector normal to $\mathfrak{M}$ in $\mathbb{R}^n_x \times \mathbb{R}^n_\xi$ (the subscripts denoting the variables in the ambient spaces) and by ν_x and ν_ξ the components of ν in the directions of $\mathbb{R}^n_x$ and $\mathbb{R}^n_\xi$. These components act on each other by the dot product. We denote also by $\delta_q(\omega)$, for a vector $\omega \in \mathbb{R}^m$, the q'th smallest component of ω. The formula for c_1 obtained in [W5] in the case $f(\lambda) = \lambda^m$ is

$$(4.1) \qquad c_1 = \frac{1}{2}(\nu_x \cdot \nu_\xi - 0i)^{1/2}\, e^{i\pi/4}\, \Gamma\left(\frac{m}{2}\right) \pi^{(m-1)/2} \sum_{q=0}^{m} \frac{\beta_{m,q}}{(2\pi)^n} \int_{\mathfrak{M}} \sigma^+(x,\xi)^{m-q}\, \sigma^-(x,\xi)^q\, d(x,\xi)$$

where $\sigma^{\pm}(x, \xi)$ denote the limits of σ from the two sides of $\mathfrak{M}$ and where $d(x, \xi)$ denotes Lebesgue measure on $\mathfrak{M}$. The coefficients $\beta_{m,q}$ are given by

(4.2) $$\beta_{m,q} = \int_{\mathcal{S}} \left[\delta_{q+1}(\omega) - \delta_q(\omega)\right] d\omega \qquad (q = 0,..., m)$$

where $\mathcal{S}$ denotes the unit sphere in the subspace of $\mathbb{R}^m$ given by $\Sigma\omega_i = 0$, and we make the convention $\delta_0(\omega) = \delta_{m+1}(\omega) = 0$.

(The notation here is slightly different than in [W5]. The validity of the expansion (0.16) was proved under the assumption that $\nu_x \cdot \nu_\xi \neq 0$. There is no doubt that it holds also when $\nu_x \cdot \nu_\xi = 0$ although a separate argument would have been needed to handle this special case.)

We are going to evaluate $\int_{\mathcal{S}} \delta_q(\omega)\, d\omega$ by relating it to

(4.3) $$\int_{\mathbb{R}^m} e^{-|x|^2}\, \delta_q(x)\, dx$$

which, in turn, we shall express in terms of the error function $E(s)$ given by (0.19).

If we write

$$\mathfrak{R} = \{x \in \mathbb{R}^m : \Sigma x_i = 0\}$$

then $\mathcal{S}$ is the unit sphere in $\mathfrak{R}$ and $\mathbb{R}^m$ is the orthogonal direct sum of $\mathfrak{R}$ and the space spanned by the unit vector $m^{-1/2}\,\overline{1}$ where $\overline{1} = (1,..., 1)$. Thus any vector in $\mathbb{R}^m$ has the representation $x + s\, m^{-\frac{1}{2}}\,\overline{1}$ with $x \in \mathfrak{R}$ and $s \in \mathbb{R}$, and (4.3) becomes

$$\int_{\mathfrak{R}} \int_{\mathbb{R}} e^{-|x|^2} e^{-s^2}\, \delta_q(x + s\, m^{-\frac{1}{2}}\,\overline{1})\, ds\, dx\, .$$

If we observe that $\delta_q(x + s\, m^{-\frac{1}{2}}\,\overline{1}) = \delta_q(x) + s\, m^{-\frac{1}{2}}$ and that $\int e^{-s^2} s\, ds = 0$ we see that the above is equal to

$$\int_{\mathfrak{R}} \int_{\mathbb{R}} e^{-|x|^2} e^{-s^2}\, \delta_q(x)\, ds\, dx = \pi^{1/2} \int_{\mathfrak{R}} e^{-|x|^2}\, \delta_q(x)\, dx.$$

Introducing spherical coordinates in the $(m-1)$-dimensional space $\mathfrak{R}$ and using the facts that $\delta_q(x)$ is homogeneous of degree 1 and that

$$\int_0^\infty e^{-r^2} r^{m-1}\, dr = \frac{1}{2}\Gamma\left(\frac{m}{2}\right)$$

we deduce that (4.3) equals

$$\frac{1}{2}\Gamma\left(\frac{m}{2}\right)\pi^{1/2} \int_{\mathcal{S}} \delta_q(\omega)\, d\omega.$$

Another way to evaluate (4.3) is to observe that it equals $m\,C^{m-1}_{q-1}$ times the integral over the region

(4.4) $$\{x \in \mathbb{R}^m : x_i < x_q \text{ when } i < q \text{ and } x_i > x_q \text{ when } i > q\}.$$

The reason for this is that there are m possibilities for the q'th smallest component of x, and for each of these there are C^{m-1}_{q-1} possibilities for the $q-1$ components which are smaller; by symmetry the integrals over the regions corresponding to all choices are equal, and the region (4.4) corresponds to a particular choice. Thus (4.3) equals

$$m\,C^{m-1}_{q-1}\int_{-\infty}^{\infty} e^{-x_q^2}x_q\,dx_q \cdot \prod_{i<q}\int_{-\infty}^{x_q} e^{-x_i^2}dx_i \cdot \prod_{i>q}\int_{x_q}^{\infty} e^{-x_i^2}dx_i\,.$$

Making the variable changes $x_i \to \pi^{1/2}s$ in all the integrals and recalling (0.18) we see that this equals

$$m\,C^{m-1}_{q-1}\;\pi^{(m+1)/2}\int_{-\infty}^{\infty} s\,E'(s)\,E(s)^{q-1}\,(1-E(s))^{m-q}\,ds\,.$$

Comparing our two expressions for (4.3) gives the desired formula

$$\int_{\mathcal{S}} \delta_q(\omega)\,d\omega = 2\,\Gamma\!\left(\frac{m}{2}\right)^{-1}\pi^{m/2}\,m\,C^{m-1}_{q-1}\int_{-\infty}^{\infty} s\,E'(s)\;E(s)^{q-1}\,(1-E(s))^{m-q}\,ds\,.$$

It follows that the constant $\beta_{m,q}$ given by (4.2) is equal to $2\,\Gamma\!\left(\frac{m}{2}\right)^{-1}\pi^{m/2}$ times

$$m\int_{-\infty}^{\infty} s\,E'(s)\,\left[C^{m-1}_{q}\,E(s)^{q}\,(1-E(s))^{m-q-1} - C^{m-1}_{q-1}\,E(s)^{q-1}\,(1-E(s))^{m-q}\right]ds\,.$$

An easy computation with the binomial coefficients shows that this is the same as

(4.5) $$C^m_q\int_{-\infty}^{\infty} s\,E'(s)\left[(m-q)\,E(s)^{q}\,(1-E(s))^{m-q-1} - q\,E(s)^{q-1}\,(1-E(s)^{m-q}\right]ds\,.$$

The expression in brackets equals

$$-\frac{d}{ds}\,E(s)^{q}\,(1-E(s)^{m-q}$$

and so if q is not equal to 0 or m integration by parts shows that (4.5) is equal to

(4.6) $$C^m_q\int_{-\infty}^{\infty} E(s)^{q}\,(1-E(s))^{m-q}\,ds\,.$$

Here we used the fact that $E(s)$ vanishes at $-\infty$ and $1-E(s)$ vanishes at $+\infty$. If q equals 0 or m integration by parts like this fails, the last integral being divergent. To see what replaces it observe that

$$\int_{-\infty}^{\infty} s\, E'(s)\, ds = 0$$

since $E'(s) = e^{-\pi s^2}$ is even. Hence when $q = m$ (4.5) is equal to

$$-\int_{-\infty}^{\infty} s\, E'(s) \left[m\, E(s)^{m-1} - 1\right] ds$$

and integration by parts shows that this equals the convergent integral

$$\int_{-\infty}^{\infty} \left(E(s)^m - E(s)\right) ds\,.$$

This is to replace (4.6) when $q = m$. Similarly its replacement when $q = 0$ is

$$\int_{-\infty}^{\infty} \left[(1 - E(s))^m - (1 - E(s))\right] ds\,.$$

Putting all this together with (4.1) and using the binomial theorem we find that in the case $f(\lambda) = \lambda^m$

$$c_1 = (\nu_x \cdot \nu_\xi - 0i)^{1/2} \frac{e^{i\pi/4}\, \pi^{1/2}}{(2\pi)^n} \int_{\mathfrak{M}} \int_{\mathbb{R}} \left\{\left[(1 - E(s))\, \sigma^+(x, \xi) + E(s)\, \sigma^-(x, \xi)\right]^m \right.$$
$$\left. - \left[(1 - E(s))\, \sigma^+(x, \xi)^m + E(s)\, \sigma^-(x, \xi)^m\right]\right\} ds\, d(x, \xi).$$

If we write $w(t) = dE^{-1}(t)/dt$, with E^{-1} denoting the inverse function of E, and set $E(s) = t$ the inner integral becomes

$$\int_0^1 \left\{\left[(1 - t)\, \sigma^+(x, \xi) + t\, \sigma^-(x, \xi)\right]^m - \left[(1 - t)\, \sigma^+(x, \xi)^m + t\, \sigma^-(x, \xi)^m\right]\right\} w(t)\, dt\,.$$

Thus, with σ^* defined by (0.16) and our integrals regularized as usual, we have shown that for polynomial f

$$c_1 = (\nu_x \cdot \nu_\xi - 0i)^{1/2} \frac{e^{i\pi/4}\, \pi^{1/2}}{(2\pi)^n} \int_{\mathfrak{M}} \int_{\mathbb{R}} f(\sigma^*(x, \xi, t))\, w(t)\, dt\, d(x, \xi).$$

REFERENCES

[B] E. Basor. *Trace formulas for Toeplitz matrices with piecewise continuous symbol.* J. Math. Anal. Applic. 120 (1986).

[BS] A. Böttcher and B. Silbermann. *Analysis of Toeplitz Operators.* Akademic-Verlag, Berlin, 1989.

[F] X. Fu. *Asymptotics of Toeplitz matrices with symbols of bounded variation.* Ph.D. Thesis, Univ. of California, Santa Cruz, 1991.

[GF] I. C. Gohberg and I. A. Feldman. *Convolution Equations and Projection Methods for Their Solution*. Amer. Math. Soc. Transl. Math. Monographs 41 (1974).

[GK] I. Gohberg and N. Krupnik. *On the algebra generated by Toeplitz matrices*. Funcl. Anal. Appl. 3 (1969).

[H] L. Hörmander. *The Analysis of Linear Partial Differential Operators III*. Springer-Verlag, Berlin, 1985.

[L] L. M. Libkind. *Asymptotics of the eigenvalues of Toeplitz forms*. Math. Notes. 11 (1972).

[W1] H. Widom. *Families of Pseudodifferential Operators*. In *Topics in Functional Analysis* (I. Gohberg and M. Kac, eds.), Academic Press, New York, 1978.

[W2] _______. *On a class of integral operators with discontinuous symbol*. J. Int. Eqs. Operator Th. 5 (1982).

[W3] _______. *A trace formula for Wiener-Hopf operators*. J. Operator Th. 8 (1982).

[W4] _______. *Asymptotic Expansion for Pseudodifferential Operators on Bounded Domains*. Lecture Notes in Math. 1152, Springer-Verlag, Berlin, 1986.

[W5] _______. *Asymptotic expansions and stationary phase for operators with discontinuous symbol*. To appear.

Department of Mathematics
University of California
Santa Cruz, CA 95064

AMS Classification Numbers:
Primary 45M05
Secondary 47B35, 47G05

UNIVERSITY OF MARYLAND AT COLLEGE PARK

DEPARTMENT OF MATHEMATICS

WORKSHOP ON CONTINUOUS AND DISCRETE FOURIER TRANSFORMS AND EXTENSION PROBLEMS

September 25 & 26, 1991

Wednesday, Sept. 25, 1991

11:00 - 11:45	Cora Sadosky	*Generalized Toeplitz kernels, weakly positive matrix measures and applications*
11:50 - 12.35	Misha Cotlar	*Two-dimensional Bochner theorem and the theorems of Adamjan-Arov-Krein and Sz.Nagy-Foias*
2:15 - 3:00	Robert Ellis	*Orthogonal systems related to infinite Hankel matrices*
3.05 - 3:40	Israel Gohberg	*Generalized Toeplitz kernels and the band extension method*
4:00 - 4:35	Israel Gohberg	*Continuous and discrete Fourier transforms, and formulas of Szegö-Kac-Achiezer*
4:40 - 5:25	Rien Kaashoek	*Connections between continuous and discrete maximum entropy principles*
5:30 - 6:50	David Lay	*Distribution of zeros of matrix-valued continuous analogues of orthogonal polynomials*

Thursday, Sept. 26, 1991

3:00 - 3:45	John Benedetto	*Continuous and discrete Fourier transforms, and sampling formulas*
3:50 - 4:35	Seymour Goldberg	*Extensions of triangular Hilbert-Schmidt operators*
4:40 - 5:25	Henry Landau	*Stability for continuous analogues of orthogonal polynomials*

Organizing Committee: R. Ellis, I. Gohberg, S. Goldberg, and D. Lay

Titles previously published in the series

OPERATOR THEORY: ADVANCES AND APPLICATIONS
BIRKHÄUSER VERLAG

1. **H. Bart, I. Gohberg, M.A. Kaashoek:** Minimal Factorization of Matrix and Operator Functions, 1979, (3-7643-1139-8)
2. **C. Apostol, R.G. Douglas, B. Sz.-Nagy, D. Voiculescu, Gr. Arsene** (Eds.): Topics in Modern Operator Theroy, 1981, (3-7643-1244-0)
3. **K. Clancey, I. Gohberg:** Factorization of Matrix Functions and Singular Integral Operators, 1981, (3-7643-1297-1)
4. **I. Gohberg** (Ed.): Toeplitz Centennial, 1982, (3-7643-1333-1)
5. **H.G. Kaper, C.G. Lekkerkerker, J. Hejtmanek:** Spectral Methods in Linear Transport Theory, 1982, (3-7643-1372-2)
6. **C. Apostol, R.G. Douglas, B. Sz-Nagy, D. Voiculescu, Gr. Arsene** (Eds.): Invariant Subspaces and Other Topics, 1982, (3-7643-1360-9)
7. **M.G. Krein:** Topics in Differential and Integral Equations and Operator Theory, 1983, (3-7643-1517-2)
8. **I. Gohberg, P. Lancaster, L. Rodman:** Matrices and Indefinite Scalar Products, 1983, (3-7643-1527-X)
9. **H. Baumgärtel, M. Wollenberg:** Mathematical Scattering Theory, 1983, (3-7643-1519-9)
10. **D. Xia:** Spectral Theory of Hyponormal Operators, 1983, (3-7643-1541-5)
11. **C. Apostol, C.M. Pearcy, B. Sz.-Nagy, D. Voiculescu, Gr. Arsene** (Eds.): Dilation Theory, Toeplitz Operators and Other Topics, 1983, (3-7643-1516-4)
12. **H. Dym, I. Gohberg** (Eds.): Topics in Operator Theory Systems and Networks, 1984, (3-7643-1550-4)
13. **G. Heinig, K. Rost:** Algebraic Methods for Toeplitz-like Matrices and Operators, 1984, (3-7643-1643-8)
14. **H. Helson, B. Sz.-Nagy, F.-H. Vasilescu, D.Voiculescu, Gr. Arsene** (Eds.): Spectral Theory of Linear Operators and Related Topics, 1984, (3-7643-1642-X)
15. **H. Baumgärtel:** Analytic Perturbation Theory for Matrices and Operators, 1984, (3-7643-1664-0)
16. **H. König:** Eigenvalue Distribution of Compact Operators, 1986, (3-7643-1755-8)

17. **R.G. Douglas, C.M. Pearcy, B. Sz.-Nagy, F.-H. Vasilescu, D. Voiculescu, Gr. Arsene** (Eds.): Advances in Invariant Subspaces and Other Results of Operator Theory, 1986, (3-7643-1763-9)
18. **I. Gohberg** (Ed.): I. Schur Methods in Operator Theory and Signal Processing, 1986, (3-7643-1776-0)
19. **H. Bart, I. Gohberg, M.A. Kaashoek** (Eds.): Operator Theory and Systems, 1986, (3-7643-1783-3)
20. **D. Amir:** Isometric characterization of Inner Product Spaces, 1986, (3-7643-1774-4)
21. **I. Gohberg, M.A. Kaashoek** (Eds.): Constructive Methods of Wiener-Hopf Factorization, 1986, (3-7643-1826-0)
22. **V.A. Marchenko:** Sturm-Liouville Operators and Applications, 1986, (3-7643-1794-9)
23. **W. Greenberg, C. van der Mee, V. Protopopescu:** Boundary Value Problems in Abstract Kinetic Theory, 1987, (3-7643-1765-5)
24. **H. Helson, B. Sz.-Nagy, F.-H. Vasilescu, D. Voiculescu, Gr. Arsene** (Eds.): Operators in Indefinite Metric Spaces, Scattering Theory and Other Topics, 1987, (3-7643-1843-0)
25. **G.S. Litvinchuk, I.M. Spitkovskii:** Factorization of Measurable Matrix Functions, 1987, (3-7643-1843-X)
26. **N.Y. Krupnik:** Banach Algebras with Symbol and Singular Integral Operators, 1987, (3-7643-1836-8)
27. **A. Bultheel:** Laurent Series and their Pade Approximation, 1987, (3-7643-1940-2)
28. **H. Helson, C.M. Pearcy, F.-H. Vasilescu, D. Voiculescu, Gr. Arsene** (Eds.): Special Classes of Linear Operators and Other Topics, 1988, (3-7643-1970-4)
29. **I. Gohberg** (Ed.): Topics in Operator Theory and Interpolation, 1988, (3-7634-1960-7)
30. **Yu.I. Lyubich:** Introduction to the Theory of Banach Representations of Groups, 1988, (3-7643-2207-1)
31. **E.M. Polishchuk:** Continual Means and Boundary Value Problems in Function Spaces, 1988, (3-7643-2217-9)
32. **I. Gohberg** (Ed.): Topics in Operator Theory. Constantin Apostol Memorial Issue, 1988, (3-7643-2232-2)
33. **I. Gohberg** (Ed.): Topics in Interplation Theory of Rational Matrix-Valued Functions, 1988, (3-7643-2233-0)
34. **I. Gohberg** (Ed.): Orthogonal Matrix-Valued Polynomials and Applications, 1988, (3-7643-2242-X)
35. **I. Gohberg, J.W. Helton, L. Rodman** (Eds.): Contributions to Operator Theory and its Applications, 1988, (3-7643-2221-7)
36. **G.R. Belitskii, Yu.I. Lyubich:** Matrix Norms and their Applications, 1988, (3-7643-2220-9)
37. **K. Schmüdgen:** Unbounded Operator Algebras and Representation Theory, 1990, (3-7643-2321-3)
38. **L. Rodman:** An Introduction to Operator Polynomials, 1989, (3-7643-2324-8)
39. **M. Martin, M. Putinar:** Lectures on Hyponormal Operators, 1989, (3-7643-2329-9)

40. **H. Dym, S. Goldberg, P. Lancaster, M.A. Kaashoek** (Eds.): The Gohberg Anniversary Collection, Volume I, 1989, (3-7643-2307-8)
41. **H. Dym, S. Goldberg, P. Lancaster, M.A. Kaashoek** (Eds.): The Gohberg Anniversary Collection, Volume II, 1989, (3-7643-2308-6)
42. **N.K. Nikolskii** (Ed.): Toeplitz Operators and Spectral Function Theory, 1989, (3-7643-2344-2)
43. **H. Helson, B. Sz.-Nagy, F.-H. Vasilescu, Gr. Arsene** (Eds.): Linear Operators in Function Spaces, 1990, (3-7643-2343-4)
44. **C. Foias, A. Frazho:** The Commutant Lifting Approach to Interpolation Problems, 1990, (3-7643-2461-9)
45. **J.A. Ball, I. Gohberg, L. Rodman:** Interpolation of Rational Matrix Functions, 1990, (3-7643-2476-7)
46. **P. Exner, H. Neidhardt** (Eds.): Order, Disorder and Chaos in Quantum Systems, 1990, (3-7643-2492-9)
47. **I. Gohberg** (Ed.): Extension and Interpolation of Linear Operators and Matrix Functions, 1990, (3-7643-2530-5)
48. **L. de Branges, I. Gohberg, J. Rovnyak** (Eds.): Topics in Operator Theory. Ernst D. Hellinger Memorial Volume, 1990, (3-7643-2532-1)
49. **I. Gohberg, S. Goldberg, M.A. Kaashoek:** Classes of Linear Operators, Volume I, 1990, (3-7643-2531-3)
50. **H. Bart, I. Gohberg, M.A. Kaashoek** (Eds.): Topics in Matrix and Operator Theory, 1991, (3-7643-2570-4)
51. **W. Greenberg, J. Polewczak** (Eds.): Modern Mathematical Methods in Transport Theory, 1991, (3-7643-2571-2)
52. **S. Prössdorf, B. Silbermann:** Numerical Analysis for Integral and Related Operator Equations, 1991, (3-7643-2620-4)
53. **I. Gohberg, N. Krupnik:** One-Dimensional Linear Singular Integral Equations, Volume I, Introduction, 1991, (3-7643-2584-4)
54. **I. Gohberg, N. Krupnik** (Eds.): One-Dimensional Linear Singular Integral Equations, 1992, (3-7643-2796-0)
55. **R.R. Akhmerov, M.I. Kamenskii, A.S. Potapov, A.E. Rodkina, B.N. Sadovskii:** Measures of Noncompactness and Condensing Operators, 1992, (3-7643-2716-2)
56. **I. Gohberg** (Ed.): Time-Variant Systems and Interpolation, 1992, (3-7643-2738-3)
57. **M. Demuth, B. Gramsch, B.W. Schulze** (Eds.): Operator Calculus and Spectral Theory, 1992, (3-7643-2792-8)